D0802504

BISON
BOOKS

CRISIS & OPPORTUNITY

Sustainability in American Agriculture

JOHN E. IKERD

University of Nebraska Press | Lincoln and London

LIBRARY OF CONGRESS
CATALOGING-IN-PUBLICATION DATA

Ikerd, John E.
Crisis and opportunity : sustainability in
American agriculture / John E. Ikerd.
    p. cm.
Includes bibliographical references.
ISBN-13: 978-0-8032-1142-1
    (pbk. : alk. paper)
    1. Sustainable agriculture—United
States. 2. Alternative agriculture—
Economic aspects—United States.
3. Farms, Small—United States.
4. Agriculture—United States—
Forecasting. I. Title.
    S494.5.S86138 2008
    630.973—dc22   2007036542

Set in Janson by Bob Reitz.
Designed by A. Shahan.

Contents

Preface

The essays in this collection have been written over the span of more than a decade. The various topics were suggested by the people who organized the sustainable agriculture conferences where they were presented. The essays address some of the most important questions of the sustainable agriculture movement: why did it begin, what is it about, and how can it succeed?

The sustainable agriculture movement emerged in response to a growing crisis in American agriculture, a crisis arising from the unintended social, ecological, and economic consequences of agricultural industrialization. Sustainable agriculture is about meeting the needs of the present without compromising the future, which requires harmony and balance among the ecological, economic, and social dimensions of agriculture. But sustainable agriculture also is about the pursuit of a desirable quality of life—materially, socially, and spiritually—rather than the pursuit of narrow individual self-interests. Finally, the sustainable agriculture movement can and will succeed as farmers, consumers, and citizens realize, one by one, that farming and living sustainably is simply a better way to farm, to work, and to live. Those who are actually farming and living more sustainably are proving that it can be done.

I have been involved with the sustainable agriculture movement since the late 1980s, when it was first publicly validated by the USDA Low Input Sustainable Agriculture program. During the decade of the 1990s, I represented the U.S. Department of Agriculture and the University of Missouri in facilitating and conducting sustainable agriculture research and educational programs all across the country. During this time, and since retiring in early 2000, I have had the great privilege of making hundreds of presentations at various public events related to sustainable agriculture. Some of my earlier presentations focused on specific issues, such as agricultural industrialization, defining sustainable agriculture, niche marketing, and sustainable community development.

In the mid-1990s I discovered an interesting pattern in the historic writings of Thomas Paine, a prominent pamphleteer during the American Revolution. He always began his pamphlets with an indictment, by stating what was wrong with the way things were. But he always went beyond the indictment to articulate his vision of how things ought to be. He then finished each pamphlet with a message of hope, stating what needed to be done to make things the way they ought to be. The essays in this book reflect this pattern: the crisis, the opportunity, and the hope for the future.

This book as a whole tells the story of sustainable agriculture in America in its varied dimensions and from a variety of perspectives. I certainly do not claim to be an expert on Canadian agriculture, but I have averaged three to four speaking engagements in Canada per year over the past ten years. While I have observed significant differences between Canadian and U.S. agriculture, I believe the current challenges and opportunities for Canadian and U.S. farmers are very much the same, particularly with respect to sustainability.

This book presents a significant number of essays, each addressing the whole of sustainable agriculture, but within a specific context and oriented toward a specific audience. Sustainable agriculture is a holistic concept; it cannot be understood by dissecting it, examining it piece by piece, and then putting the pieces back together. A sustainable agriculture is a living system; it is individualistic, site-specific, and dynamic. Thus, sustainability must be assessed within a specific context of people, place, and time. Sustainable agriculture is diverse, and thus cannot be captured in a few examples or studies. Together, these essays tell a single holistic story of a dynamic and diverse sustainable agriculture.

The story begins with the essay "Crisis and Opportunity in American Agriculture," which establishes the theme and the pattern for the other essays and for the book as a whole. After reading the first chapter, readers should be able to skip to any section or any chapter of the book, if they prefer. Each essay stands on its own as a whole within a whole, although some sections of some essays have been edited and condensed to minimize duplication. The essays are organized in a logical progression and are mutually supportive or interdependent, but no essay is necessarily dependent upon another.

Each of the book's five sections contains three or more essays that share common themes. The first section focuses on the crisis, the industrialization of agriculture. The next two sections address the opportunity, first in general and then more specifically in terms of sustainable agriculture. The final two sections outline a new vision of hope for the future. The fourth section focuses on the hope for sustainable farms and rural communities and the last section emphasizes the hope for a sustainable food system and a sustainable society.

The knowledge and learning reflected in this book was acquired in large part from my interaction with farmers, consumers, and interested citizens as I traveled back and forth across North America. The knowledge acquired at each venue contributed to a continually evolving paradigm of sustainable agriculture, as each new presentation provided an opportunity to test new propositions or hypotheses. Over time, the ideas that worked—that were logical, relevant, internally consistent, and capable of being communicated—were added to a growing understanding of sustainable agriculture. Ideas that were not internally consistent, relevant, or grounded in reality were quickly challenged, reexamined, and either revised or discarded. The story of sustainability is a continually evolving story.

Some readers will appreciate the indigenous knowledge, evolving paradigm, and farmer-tested ideas upon which these essays were based. Others may value the book more as a source of inspiration and hope than a source of new information or knowledge.

A growing number of people understand that social and spiritual values cannot be omitted from the study or practice of sustainability. Sustainable agriculture is rooted in science, but it is equally rooted in values and ethics. I hope this book encourages other scholars to integrate science, values, and ethics and to share the resulting knowledge with others. The sustainable agriculture movement is perhaps most important because it is guiding us all toward a better way of life.

Acknowledgments

Credit for the ideas expressed in this book must be shared with the many farmers and other participants at the conferences where I have spoken over the years. I sometimes tell people that all I do is share what I have learned from farmers with other farmers, learning as much as I teach as I go from place to place. There is much truth in this statement.

Perhaps I have a more general conceptual framework upon which to hang the bits of information and knowledge that I glean, but without the real-world experiences of others to validate my conclusions, the concepts would be of little practical value. I also learn from nonfarmers and other presenters, as I try to attend as many sessions as I can at the venues where I speak. The contributions of all these people have been invaluable.

I also want to express my appreciation to my wife, Ellen, who has served as a sounding board for the concepts and expressions in these essays. She listens and reacts to my evolving thoughts and helps move the process forward, in addition to proofing every paper that I write. Together, I hope we have been able to convey a message of hope for the future of American agriculture and for the whole of humanity.

Crisis and Opportunity

1

Crisis and Opportunity in American Agriculture

North American agriculture is in crisis. Until recently, the crisis had been a quiet one. No one wanted to talk about it. Thousands of farm families were being forced off the land each year, but we were being told by the agricultural establishment that their exodus was inevitable—in fact, it was a sign of progress.[1] Those who failed were simply the victims of their own inefficiency, their inability to keep up with changing times, their inability to compete. We have no more reason to be concerned about the demise of the family farm than we were about the mom-and-pop grocery story or the family-owned restaurant. We can't stand in the way of progress, they said.

With farm prices at or near record low levels for 1997, 1998, and 1999, even the agricultural establishment began to realize that something was wrong. The U.S. Congress passed emergency farm legislation each of those three years, pushing U.S. farm subsidies to all-time record levels. But even then, the farm crisis was being blamed on such things as weather problems, loss of export markets, or unwise public policies.

In general, we are led to believe that our farm problems are someone else's fault. The crisis is a simple matter of supply and demand, we are told. The only solutions being seriously proposed are to tinker with government policy, or better yet, to simply wait for markets to recover. In the meantime, the only alternatives farmers are being offered are to get big enough to be competitive, get a corporate contract to reduce risks, or get out of farming.

Eventually, prices for agricultural commodities will recover, at least for a year or two. Weather problems in a major exporting country will tighten global supplies, a crop failure in a major importing country will spark global demand, or changes in financial markets will shift global trade patterns. Agricultural markets are inherently unstable. However, a year or two of profitable prices will do nothing to resolve the underlying problems of American agriculture.

In a recent book, *The End of Agriculture in the American Portfolio*, University of California economist Steven Blank envisions the imminent end of the American farm.[2] His conclusions regarding agriculture in the United States would seem to be equally applicable to agriculture in Canada. American agriculture is coming to an end, he argues, but he claims this should be no cause for alarm. He contends that the end of agriculture in America is the result of a natural process that is making us all better off. He foresees a time in the not too distant future when North America will import nearly all its foodstuffs from other, "lesser developed" countries. Costs of land and labor will be too high for American farmers to compete in global commodity markets. He argues that globalization of the food system is not some corporate conspiracy but is simply the inevitable consequence of the individual struggles of farmers and agribusiness in America and around the world who quite logically are pursuing their individual self-interests, which ultimately will benefit society in general.

Blank believes that the current open spaces of rural areas will be transformed from farms to residential developments to accommodate a growing and increasingly affluent population fleeing the problems of urbanization. Cornfields will be unable to compete with condominiums for farmland. Farming is a low-skilled *primary* industry that has no place in an advanced high-tech economy. Rural ways of life will give way to urban ways of life, as farms become residential ranchettes. Virtual communities of people interconnected by the Internet will replace real communities of people who meet face to face in church or at the grocery store. Communities of interest will replace communities of place. Agriculture will no longer be a significant factor in the rural economy. Most people in the community will be employed elsewhere—perhaps by companies thousands of miles away. Blank claims the only forms of truly sustainable agriculture will be those compatible with urban life—mainly golf courses, plant nurseries, and turf farms.

Blank's fundamental arguments are based on the premise that economic considerations ultimately will prevail over all others. He assumes that industrial agribusinesses will replace family farms because they are more economically efficient and that American agribusiness eventually will be displaced by even more efficient agribusiness elsewhere in the global economy. Residential ranchettes will replace rural farmsteads because people with high-tech jobs can pay more for land to look at than farm families can afford to pay for land to work on.

Blank might well be right, if we allow short-run economic thinking to continue to dominate every aspect of our lives. The current crisis in agriculture might well foretell the end of North American agriculture. However, the end of farming in North America is neither inevitable nor desirable. There are sound logical, ecological, and social reasons to keep farm families on the land and for every nation to maintain the integrity of its agricultural sector. We need not sacrifice our national food security and our quality of life for the sake of short-run economic efficiency. But we may be forced to rethink the role and scope of agriculture within the global economy, as well as within human society. We may have to develop a new American farm to prevent the end of the American farm.

American agriculture is at a time of crisis. Crisis is most frequently considered something negative, something to be avoided, such as pain, distress, or disorder. However, crisis can be defined more generally to be either positive or negative. A crisis is a decisive moment, a critical time, or state of affairs whose outcome will make a decisive difference for either better or worse.[3]

The current crisis in agriculture most certainly is a time of pain, distress, and disorder for farmers and rural communities. However, it is also a time of opportunity—a critical time and state of affairs that will make a decisive difference, either for better or for worse. Rather than passively accept whatever might happen, it's up to us—to farmers and others—to confront the threat, seize the opportunity, and create the kind of agriculture and human society that we want.

I. Disagree ⟹

To seize the opportunity, we first must be willing to confront the crisis. The current crisis in agriculture is not a consequence of the weather, world trade problems, or unwise government policies. These things only magnify the symptoms of problems that are rooted in causes far more fundamental. Crisis is a chronic symptom of the type of agriculture we have been promoting on this continent for at least the past fifty years—symptoms of an industrial agriculture. Reoccurring financial crises are the consequence of our encouraging farmers to industrialize—to become more specialized, standardized, and larger in scale to make agriculture more efficient. We rationalize the industrialization of agriculture as a necessary means of providing lower-cost food for consumers. We rationalize the displacement of family farmers in the process as a necessary means of "freeing people from the drudgery of farming" so they can find better jobs elsewhere. We are led to believe that the benefits far outweigh any costs.

The promise of profits is the bait that keeps farmers on the treadmill of industrialization. Farmers adopt new cost-cutting and production-enhancing technologies to increase profits, but the resulting increases in production cause prices to fall, eliminating the profits of early adopters and driving the laggards out of business. This technology treadmill has been driving farmers off the land for decades. Those remaining on the treadmill after each crisis must run faster and faster to survive. Soon they don't have enough time for their families, let alone their communities. They can't afford to care too much about their neighbor because they know they will soon have to have their neighbor's land in order to survive. Inefficiency and reluctance to change are not the causes of failure among American farmers. Failure is an inherent part of the current system of farming. Some farmers must fail in order for others to succeed, and after each crisis, there is room for fewer survivors.

Chronic crisis in American agriculture also has meant chronic crisis in America's rural communities, as farms have become more specialized, larger, and fewer. The fundamental purpose of most rural communities was to support those engaged in agriculture, and thus the communities were supported by those involved in agriculture. It takes people, not just production, to support a community. People buy automobiles, appliances, clothes, and haircuts on the main streets of farming towns. Larger farms tend to bypass rural communities when buying their production inputs and marketing their products. In addition, a rural community is far more than a rural economy. It takes people to fill the church pews and school desks, to serve on town councils, to justify investments in health care and other social services, to do the things that make a community. As farms have grown larger and fewer, rural communities have lost people—human and social resources—and many rural communities have withered and died.

However, the current crisis is different from others in at least one respect: it signals the final stage of industrialization. The final stage is consolidation of decision making under corporate control. The giant multinational corporations are now seizing control of all aspects of American agriculture, moving beyond specialization and standardization, beyond consolidation into larger farms, and are now consolidating agricultural decision making into the boardrooms of a handful of multinational corporations. This final stage of industrialization is turning once peaceful farms into odious factories, with all the noxious odors, environmental degradation, and inhumane working conditions that characterized heavy industry of earlier times. This final stage of industrialization is turning remote rural communities into the dumping grounds for the rest of society—whether for prisons, landfills, toxic waste dumps, or giant confinement animal factories. This final stage of industrialization could well spell the end of the American farm, and with it, the end of the American rural community.

The corporatization of agriculture in the United States came first to the poultry industry. A handful of larger corporations now control poultry production from genetics to the supermarket, and there are virtually no independent producers left. Hog production is rapidly following in the footsteps of poultry, with corporate ownership and contract production becoming the norm rather than the exception. Dairy will likely be the next sector to industrialize, as the current trend toward large-scale production will quite likely be followed by corporate control, or corporate-like cooperative control. Biotechnology will bring corporate control of grain production, as genetic engineering is used to create specific characteristics of food products made from grain. Producers will then have to grow crops with approved genetics in order to have a market, and biotech corporations will hold the genetic patents. A grain farmer who doesn't sign a corporate contract simply won't have a market.

It's not a matter of economies of scale any longer but instead a matter of market control. Market control translates into profits. Poultry producers have proven that if a few corporations can gain control of a sufficiently large share of an industry, they can stabilize supplies on the backs of their contract producers and can maintain corporate profits indefinitely. During the consolidation phase, however, corporate producers are not concerned with maximizing profits. The lower the price, the faster independent producers will be forced out of business and the faster the large corporations can gain market share. As corporations gain market share, they can deny market access to lower-cost independent producers and ultimately gain control of the market, even if they are less efficient than are independent producers. When they get control of the markets, they can quickly recoup any losses incurred during the period of consolidation.

As American agriculture comes under corporate control, it will respond even more quickly to global markets; multinational corporations have no sentimental attachment to any particular farm, geographic region, or nation. If costs of land and labor are less somewhere other than in North America, as they almost certainly will be, then that's where our food will be produced. Capital and management can be shifted easily from North America to other regions around the globe, as we have seen in the production of other industrial goods. North America's farmlands will be sold to the highest bidder, which is likely to be land speculators, and most rural communities will continue to wither and die as they await some future economic revival such as becoming bedroom communities for affluent urbanites.

The food and fiber industry most certainly has a future. People will always need food, clothing, and shelter, and someone will provide these things. But there will be no future for farming in North America, or for rural farming communities, unless we challenge the conventional wisdom that food should be produced wherever on the globe it can be produced at the lowest cost and that free markets should be the final arbitrators of all value. In fact, there will be no future for farming anywhere—not true farming—unless we find the courage to challenge and disprove the conventional wisdom that farmers must get bigger, give in to corporate control, or get out. There are better alternatives for farmers and for society if we can find the courage to challenge the basic forces driving the corporatization of agriculture and of North American society.

Thankfully, the crisis in agriculture also brings with it opportunities for decisive, positive change. The opportunities arise from the failures of corporate industrialization. Economists argue that cost-reducing technologies and the pursuit of profits ensure that consumers get the highest-quality food at the lowest cost, even if some farmers are forced out of business in the process. However, we no longer have a competitive, capitalistic economic system to ensure that new technologies actually benefit consumers or that lower production costs are translated into lower food costs in the supermarket. Economists are defending corporate agriculture using hopelessly outdated theories developed more than two hundred years ago, during completely different times.

Contemporary economics is based on the observations of British economist Adam Smith in his landmark book, *The Wealth of Nations*, published in 1776. From Smith's observations, economists developed the fundamental assumptions or conditions that underlie free market economic thinking even today.[4] These conditions must hold in order for Smith's "invisible hand" of competition to transform individual greed into the greater good for society in general.

Markets must be economically competitive—meaning the numbers of buyers and sellers are so large that no single buyer or seller can have any noticeable effect on market price. In such markets, cost savings are quickly passed on to consumers and no one in the system has the power to exploit anyone else. It must be easy for new sellers to enter markets that are profitable and easy for sellers to get out of unprofitable markets, so that producers are able to respond to consumers' changing wants and needs. Consumers must have clear and accurate information concerning whether the things they buy will actually meet their wants and needs. And finally, the consumer must be sovereign—they must be free to choose according to tastes and preferences reflecting their basic values, untainted by persuasive advertising or influences of others.

None of these conditions exist in today's society. Today, agricultural markets are dominated by a few large agribusiness corporations, certainly at every level other than that of farming, and increasingly even at the farm level. In addition, it is not easy to get into or out of any aspect of agribusiness, and it is becoming increasingly difficult even to get into or out of farming. Consumers often don't get accurate, unbiased information concerning the products they buy; instead, they get disinformation by design, disguised as advertising. Finally, consumers are no longer sovereign. The food industry spends billions of dollars on advertising designed to bend and shape consumers' tastes and preferences to accommodate mass production and mass distribution, which enables corporate control. There is no logical reason to believe that the corporate agriculture of today is evolving to meet the needs or wants of consumers.

Instead, corporate agriculture is evolving to generate more profits and growth for the benefit of corporate investors, at the expense of both consumers and farmers. We no longer have a competitive, capitalistic agricultural economy. Capitalism requires that individuals make individual decisions in a competitive market environment. As corporations extend their control horizontally *within* the same functional levels, such as marketing, storage, transportation, processing, or retailing, they increase their ability to protect their profits from competitors. As corporations also extend their control vertically, *across* functional levels, including additional, different stages of production and marketing, they gain control over decisions concerning how much of a product is produced, when it is produced, how it is produced, and for whom. The corporations make decisions designed to maximize their profits and growth, not to meet the needs of society.

In essence, as agriculture moves from competitive capitalism to corporatism, it changes from a market economy to a centrally planned economy. Central planning didn't work for the communists and it won't work for the corporations. The problem with communism was not that communists weren't smart enough or that their computers weren't large enough. Central planning is a fundamentally wrong-headed approach to managing an economy—for corporations as well as for governments. The corporate system of food production will prove to be fundamentally incapable of meeting the needs of the people. A prime opportunity now exists to create new systems of food and farming that will truly meet the needs of the people of an enlightened society. In spite of continued exploitation by corporations, our society is slowly becoming more enlightened. And as a result, we have begun to realize we are destroying our natural environment in the process of trying to produce cheap food. We are mining the soil through erosion and depletion of its natural productivity in the process of maximizing production and minimizing dollar-and-cent costs of production. We are polluting our streams and groundwater with residues from the pesticides and commercial fertilizers necessary for large-scale specialized crop production and with wastes from giant confinement animal feeding factories. We are destroying the genetic diversity that is necessary to support nature's means of capturing and transforming solar energy into energy to sustain human bodies.

As society becomes more enlightened, we are also beginning to realize that we are destroying the social fabric of society in the process of trying to make agriculture more efficient. We are destroying opportunities for people to lead productive, successful lives. We are turning thinking, innovative, creative farmers into tractor drivers and hog-house janitors. There is dignity in all types of work, but all people should have opportunities to express their full human potential. Consolidation of decision making concentrates the opportunities among the privileged few while leaving the many less fortunate without hope for a rewarding future. Industrial specialization also tends to separate people within families, within communities, and within nations. We are just beginning to realize that industrialization destroys the human relationships needed to support a civilized society.

The outdated economic theory that supports agricultural industrialization is fundamentally incapable of dealing effectively with either the environmental or the social challenges of today. In economics, the environment and society are considered to be external to the decision-making process—that is, they may either impact or be impacted by economic decisions but are not considered part of the process. In reality, the economy, environment, and society all are parts of the same inseparable whole. As society becomes more enlightened, we are beginning to see the need for a more enlightened system of decision making—one that is capable of integrating economic, ecological, and social decisions. We need a new approach to farming in North America.

Luckily, a new American agriculture already is emerging under the conceptual umbrella of sustainable agriculture. Sustainable agriculture is a response to the growing awareness that an agriculture that degrades the natural environment and weakens the social fabric of society cannot meet the needs of people over time, no matter how productive and profitable it may appear to be in the short run. Farm profitability cannot be sustained unless farms also are ecologically and socially sustainable. The focus of agricultural sustainability is on the long run, on intergenerational equity. A sustainable agriculture must be capable of meeting the needs of the present while leaving equal or better opportunities for the future.

In order to fulfill this purpose, a sustainable agriculture must be ecologically sound and socially responsible as well as economically viable. To sustain its productivity, agriculture must conserve and protect the natural resources, including the land, upon which it ultimately depends. If agriculture is to be sustained by society, it must meet the needs of society, not just as consumers but also as producers and citizens of an equitable and just society. And finally, a sustainable agriculture must be economically viable, because if all the ecologically sound and socially responsible producers go broke, then agriculture obviously will not be sustainable. Systems of farming that are lacking in any one of these dimensions quite simply are not sustainable.

Farming sustainably is no simple task, but thousands of farmers are finding ways to create a desirable quality of life for themselves and to support their local communities while being good stewards of the land and the natural environment. These farmers, like people in general, are pursuing their self-interest. Pursuit of self-interest is an inherent aspect of being human. However, people, by nature, do not pursue only their narrow, individual, or personal self-interest. It's also within the inherent nature of people to care about other people and to care for the earth. People are perfectly capable of rising above selfishness and greed to pursue a higher concept of self-interest, one that values relationships with other people and stewardship of the earth as important dimensions of one's self-interest.

This more enlightened concept of self-interest includes our narrow self-interest, which is individual and personal, but it also includes interpersonal interests, which we share with others in families and communities, and interests that are purely altruistic or ethical, which benefit others whom we never expect to know. All three contribute to our overall well-being or quality of life, explicitly recognizing that each of us individually is but a part of the whole of society, which in turn must conform to some higher order of natural law.

Sustainable agriculture requires that farmers find balance and harmony among the economic, social, and ecological dimensions of their farming operations—among self-interests, shared-interests, and purely altruistic interests. By pursuing their enlightened self-interest, these new American farmers are not only helping to build a more sustainable agriculture but are defining the principles of a more sustainable human society.

These sustainable farmers may carry the label of organic, low-input, alternative, biodynamic, holistic, permaculture, or no label at all, but they are all pursuing a common economic, ecological, and social goal. These farmers, not the experts or the scientists, are the ones on the new frontier—the explorers, the colonists, the revolutionaries, and the builders of a new world. Life is difficult on this new frontier because no one really knows how to do what these folks are trying to do; they are creating the future. They will continue to confront hardships and frustrations and there will be some failures along the way. But more and more of these new American farmers are finding ways to succeed.[5]

There are no blueprints for the new American farm, but a few fundamental principles are beginning to emerge. New American farmers focus on working with nature rather than against it. Industrial farming systems have had to bend nature—to augment, supplement, alter, and force it—to create an illusion of conformity out of nature's diversity in order to meet the demands of large-scale industrial production. The ecological problems arising from industrialization are symptoms of natural resources being used in ways that inherently degrade their natural productivity. Thus, industrialization has created tremendous opportunities for those farmers who learn to utilize the inherently productive capacity of a diverse natural-resource base, rather than wasting time and money trying to force nature to conform.

These new American farmers utilize practices such as management-intensive grazing, integrated crop and livestock farming, diverse crop rotations, cover crops, and intercropping. They manage their land and labor resources to harvest solar energy, utilizing the productivity of nature, and thus are able to reduce their reliance on external, purchased inputs. They are able to reduce costs and increase profits while protecting the natural environment and supporting their local communities.

The new American farmers focus on creating value. They realize that each of us values things differently because we have different needs and different tastes and preferences. Industrial methods are efficient only if large numbers of us are willing to settle for the same basic goods and services, to facilitate mass production. Industrialization has to treat us as if we're all pretty much the same. Customers have to be persuaded, coerced, and bribed to buy the same basic things rather than the things they really want. So we end up paying more for packaging and advertising of food than we pay to the farmers who produce the food. The industrial system creates tremendous untapped opportunities for farmers who can tailor their products to conform to unique needs and preferences of individual customers, rather than try to bend the preferences of customers to conform to their products.

The new American farmers market in the niches. They market directly to customers through farmers' markets, roadside stands, community-supported agriculture (CSAS), home delivery, or by customer pickup at the farm. They use everything from the Internet to word of mouth to make contact with their customers. They market to people who care where their food comes from and how it is produced—locally grown, organic, humanely raised, hormone-and antibiotic-free, and so on. They are often able to avoid some or all of the processing, transportation, packaging, and marketing costs that make up 80 percent of the total cost of mass-marketed foods. They increase value, reduce costs, and increase profits while protecting the environment and helping to build stronger local communities.

New American farmers focus on what they can do best. They realize that we are all different—as producers as well as consumers. We have widely diverse skills, abilities, and aptitudes. Industrialization has had to bend people—train, bribe, and coerce them—to make them behave as coordinated parts of one big machine rather than as fundamentally different human beings. Many social problems of today are symptoms of people being used by industrial systems in ways that inherently degrade our uniquely human productive capacities. Thus, industrialization has left tremendous untapped economic opportunities for farmers and others who can use their unique capacities to be productive rather than attempt to conform to systems of production that just don't fit them.

The new American farmers may produce grass-finished beef,

pastured pork, free-range or pastured poultry, heirloom varieties of fruits and vegetables, dairy or milk goats, edible flowers, decorative gourds, or dozens of other products that many label as agricultural "alternatives." They find markets for the things they want to grow and are able to grow well rather than produce for markets in which they can't compete. Or they may produce fairly common commodities by means that are uniquely suited to their talents. Their products are better, their costs are less, and their life is better because they are doing the things that they do best.

These new American farmers focus on creating value through building unique relationships—among consumers and producers and with nature. In general, they link people with purpose and place. By linking their unique productive capacities with unique sets of natural resources to serve the needs and wants of unique groups of customers, they create unique systems of meeting human needs. Uniqueness cannot be industrialized. The farmers and their customers are not just sellers and buyers, they know and care about each other as people; they have personal relationships with each other. The land is not just a resource to be exploited for economic gain. Farmers also care about and want to take care of the land; they have personal relationships with the land. The greater the uniqueness of combinations of person, purpose, and place, the more valuable will be their relationships and the more sustainable will be the value. The sameness of industrialization creates opportunities for unique farmers who can create unique relationships with their resources and their customers.

Critics argue that these new farm opportunities are limited. On the contrary, there are no limits to the diversity among people or diversity within nature. There are as many niche markets as there are different people and places. In a sense, all consumer markets are niche markets. The question is not whether there can be enough niches, but instead, how many different niches can logically be served separately. Likewise, there are as many differences in production capabilities as there are producers and as many different niches in nature as there are fields or places to produce.

Some question whether a sufficient number of people can be found who are both willing and able to learn to farm in these new ways. Admittedly, the new American farm will require a lot more knowledge, understanding, and thinking than does farming by industrial methods. However, any future occupation offering an opportunity for a decent living will require that people use their minds. The days when a person could earn a good living by just the sweat of his or her brow are in the past. There will be plenty of innovative, creative, hard-working people to operate the new American farms, once the real possibility for a more desirable quality of life in farming—economically, socially, and ethically—becomes widely known.

Others question whether people can afford to pay farmers the full costs of meeting their food and fiber needs without exploiting either the natural or human resource base of agriculture. Today's consumers, on average, spend only a dime of each dollar of their disposable income for food—from which the farmer only gets about two cents. Eight cents of the dime goes for processing, transportation, advertising, and other marketing services.[6] It would take an increase in farm-level costs of 50 percent, for example, to add even a penny to each dime consumers spend for food in the supermarket. Thus, most consumers can afford to pay farmers to produce the food they really want and need rather than settle for something less, particularly if that something less degrades the social and ecological systems from which consumers also derive much of their quality of life.

Some question whether a sustainable agriculture is physically capable of meeting the needs of a growing global population, contending that "high-yield, high-input" systems are necessary to keep pace with population growth.[7] First, "high-yield" systems rely on high levels of nonrenewable inputs such as commercial fertilizers and pesticides. Biotechnology will not reduce this reliance but instead may even increase use of pesticides and fertilizers in the quest for maximum yields. There might be sufficient supplies of these nonrenewable agricultural inputs for another fifty years, as the advocates of high-input farming claim. But what will people do for food then? We probably will have 50 percent more people on earth by then and the critical nonrenewable resources will be gone.

Many "low-input" farmers today are already achieving yields equal to or greater than conventional high-input systems of farming. The knowledge and expertise required to achieve high yields with low inputs are not nearly as common among farmers as are commercial agricultural technologies. However, many others are capable of acquiring this ability, if they realized it was possible and had an incentive to do so. In addition, sustainable agriculture today is in its infancy; sustainable farmers are but the early explorers on a new frontier. As they accumulate increased understanding and know-how, their productive abilities will undoubtedly increase as well. If we invest a fraction of the research and development efforts on regenerative farming methods in the future that we have invested in industrial methods in the past, our overall ability to produce by sustainable methods in the future may easily surpass our ability to produce by conventional methods.

Over time, as more farmers gain a better understanding of sustainable farming, productivity will rise and costs of production will fall for sustainable systems. Over time, as costs of nonrenewable inputs rise and the natural environment is further degraded, productivity will fall and costs of production will rise for industrial systems. Over time, sustainable systems will become far more productive and far less costly than industrial systems of farming.

Those who think that we can't meet the legitimate food and fiber needs of humanity with a sustainable agriculture are the "new Malthusians." Some two hundred years ago an economist by the name of Thomas Malthus claimed that humanity was destined to starve to death because population increases geometrically and technology only increases arithmetically.[8] Malthus was wrong because he failed to appreciate the potential productivity of the human mind. Those who think we can't feed the world without destroying the natural environment and without degrading human society, like Malthus, are failing to appreciate the potential role of human creativity and ingenuity in developing more sustainable systems of farming. The perceived limits to sustainable farming arise from economic assumptions that are hopelessly out of date and an industrial mindset that is rapidly losing its relevance to reality.

It's only reasonable for farmers and others to be skeptical about whether farming in general actually can be reshaped by the principles of sustainability. After all, farming is only a small part of the economy, the economy is only one aspect of human society, and industrialization has been ingrained into human society for more than two hundred years. Change may not come quickly and it may not come easily, but change will come. American agriculture fifty years from now will be fundamentally different from agriculture today; the question is not if, but how. The challenge is to change it in ways that will better serve the long-term needs of people—consumers, farmers, rural residents, and society in general—rather than the short-run economic needs of corporations. The challenge is to develop an agriculture that is ecologically sound and socially responsible so that it can also be economically viable.

To meet this challenge, we will need to have the courage to challenge the conventional wisdom that whatever is dictated by short-run economic self-interest is inevitable and is inevitably good for society. It is not. We as individuals need not wait for society to change before we can change our own lives, including our work. But we do need to overcome the economic misperceptions in our minds. We need to call on our common sense to inform us that money isn't everything; our relationships with other people matter, as does our stewardship of the natural environment. Our lives will be better when we live with harmony and balance among the personal, interpersonal, and spiritual dimensions of our lives. We must be willing to rethink what we want out of life. We have the power to enhance our quality of life, and we can begin using that power today.

For farmers, finding harmony and balance may mean changing, in very fundamental ways, the ways they farm. One of the most common stories among the new American farmers is of those who were once conventional farmers, on the technology treadmill, farming more and more land, with bigger and bigger equipment, going farther and farther in debt. Many were the "winners" in the continuing struggle for survival but found their quality of life sinking lower with each round that they "won." They didn't have the time or energy to maintain positive relationships with their spouse or their children, and they didn't have the time or economic freedom to take care of their land. They had to put all their time, energy, and money into growing the farm.

But one day these farmers realized that what they were doing didn't make sense. The more they produced and the more money they earned, the more miserable they became. And then, as many have said, they decided to cut back on the amount of land they farmed—they decided "to go back to farming the old home place but to farm it differently" rather than try to farm the whole countryside. They put their imagination and creativity into finding ways to farm that would enhance their overall quality of life—socially, spiritually, and economically—instead of focusing all their attention on production and profits. As a consequence of pursuing a higher quality of life through harmony and balance, they have developed more sustainable systems of farming and a better way of living. The world around them may have remained the same, but their world has changed. We all have the power to make the same kinds of changes in our lives.

In addition to changing our personal world, we can at least

influence a small part of the world around us. A farmer can make a difference in the land on his or her farm and in the land of others downstream. A farmer can make a difference in the lives of his or her customers and neighbors. We can all have an influence on the other people in our families, in our places of work, or in our communities. As we change our own lives in positive ways, we begin to influence those who share our little piece of the world. One by one, as we influence our little pieces of the world, the world begins to change.

We will also be more effective when we go into the public arena to advocate larger social and political change, because we will be coming from a position of self-confidence rather than desperation. We can advocate changes that are good for the whole of society over the long run, rather than support policies that might put money in our pockets at the expense of someone else. As we raise our standards in the public arena, we may find that others feel compelled to raise their standards as well.

One by one, as we find the courage to demand something better, we will change the world for the better. Susan B. Anthony, the champion of voting rights for women in the United States, once said, "Cautious, careful people, always casting about to maintain their reputation and social standing, never can bring about reform. Those who are really in earnest must be willing to be anything or nothing in the world's estimation."

It takes courage to bring about change. But as each of us finds the courage to change ourselves, we will begin to change the world. We can confront the crisis in agriculture; we can help make the outcome better rather than worse. We just need to find the courage to pursue the opportunities of sustainability.

⁝ Presented at Recapturing Wealth on the Canadian Prairies, a farming conference sponsored by the *Manitoba Co-Operator*, Brandon, Manitoba, Canada, October 26–27, 2000.

PART ONE

The Industrialization of American Agriculture

2

Why We Should Stop Promoting Industrial Agriculture

I always appreciate an opportunity to speak at the Breimyer Seminar, regardless of my topic. I told the conference organizers they could give my presentation any name they wanted this year and I would try to deal with it. The title they chose was "Why I Don't Like Industrialization and Want It Stopped." I'm sure that title was meant to be provocative, to spark some debate of the issue of agricultural industrialization. I have decided to change the title to make it bit more academic but hopefully not any less provocative. What I do or don't like about industrialization, and whether I personally do or don't want it stopped, is not of any particular significance. I have personal opinions on those matters, but they are no more important than yours or anyone else's. So rather than focus on my opinions, I intend to rely on the science of economics and on logic to make an objective case against the continued industrialization of agriculture.

The title I will actually address is "Why We Should Stop Promoting Industrial Agriculture." I will address the fundamental economic and social motives for the industrialization of agriculture, because there are sound, logical reasons *for* industrialization. However, there are also sound, logical reasons to question industrialization. As a public-sector scientist—working for the taxpayers—it's not my responsibility to stop the industrialization of agriculture. That decision is up to the people. However, I do have a responsibility to question whether we should be using public dollars to promote it. Our job is to provide people with objective information, not to promote anything other than the pursuit of truth. The people must decide what they want to stop or promote, based on that information.

I have three basic reasons for questioning the industrialization of agriculture. First, we have already realized virtually all the potential economic and social gains from the industrialization of agriculture. Those gains were significant, but there simply is very little left to be gained from further industrialization—from further specialization, standardization, and consolidation of agricultural production and marketing. Second, there are rising costs—environmental, social, and economic costs—associated with the industrialization process. In fact, the total marginal costs of industrialization may have exceeded its marginal benefits as far back as the 1970s or even 1960s. Third, and as a consequence of the other two, there is growing evidence that the industrial era is coming to an end, as it has already ended in many sectors of our economy. Industrialization was the model, or paradigm, for human progress in the twentieth century, but as we approach the twenty-first century, it is rapidly becoming obsolete. We should focus our scarce public resources on exploring approaches that have possibilities for progress in the century ahead rather than on promoting a model whose century has passed.

Peter Drucker, a noted and time-honored consultant of twentieth-century industrial managers, discusses the transformation from an industrial to a postindustrial society in his book *Post-Capitalist Society*. He states, "Every few hundred years in Western history there occurs a sharp transformation. Within a few short decades, society rearranges itself—its worldview; its basic values; its social and political structure; its arts; its key institutions. Fifty years later, there is a new world. . . . We are currently living through just such a transformation."[1]

In the late 1800s, as we began to industrialize agriculture, the potential gains from continuing the industrial revolution in agriculture were undeniable. At that time, we were still an agrarian society. More than half the people of this country were either farmers or residents of rural communities, and it took about half our total resources—money, time, and effort—just to feed and clothe ourselves. If we as a nation were to realize the emerging opportunities of the industrial revolution—to become the modern society that we know today—we had to do two things.

First, we had to free people from the task of farming so they could go to work in the factories and offices of the emerging industrial economy. Second, we had to free up some of the income and other resources being spent on food and clothing so people could buy the things these new industries were going to produce. In short, we had to make American agriculture more efficient. We had to make it possible for fewer farmers to feed more people better at a lower real cost.

The industrialization of agriculture allowed us to accomplish those two things. Through specialization, standardization, and consolidation of control, we bent nature to serve our material needs. We gradually harnessed the vagaries of biological processes and transformed farms into factories without roofs. Our fields and feedlots became biological assembly lines with production inputs coming in and agricultural commodities going out. We achieved the economies of large-scale specialized production as we applied the principles, strategies, and technologies of industrialization to farming.

Publicly funded research and education programs supplied many of these new industrial technologies and strategies. New technologies reduced per unit production costs and thus gave farmers a built-in profit incentive for their adoption. The promise of profits was even greater for those farmers who expanded production. But as more farmers increased production, market prices soon dropped by as much or more than the reduction in costs. The only benefit that later adopters realized was an opportunity to continue farming, at least for a while, and those who adopted too little too late were "freed" from farming to go to work in the factories.

This industrialization of American agriculture resulted in the most efficient agriculture in the world, at least in terms of the dollar-and-cent costs of production. A more efficient agriculture made it possible for this nation to build the strongest economy and the most affluent society in the world. The agricultural sector can be proud of these past successes. But now the objectives of industrialization have been achieved. Most of the benefits that industrialization could bring to America already have been realized.

Today, less than 2 percent of the people in this country are farmers and about half of those consider something other than farming as their primary occupation.[2] Today as a nation, we spend only about 10 percent of our disposable income for food.[3] Equally important, farmers get to keep less than 10 percent of the total amount spent for food. Eighty percent goes to marketing firms for processing, transportation, advertising, and other marketing services, and more than 10 percent—more than half of what farmers receive—goes to agribusinesses to pay for fertilizer, machinery, fuel, and other agricultural inputs.[4]

Any future gains from the further industrialization of agriculture, from improving the economic efficiency of farming, must be squeezed out of the farmer's share. And there just isn't much that can possibly be squeezed out of what little is left in farming. It simply doesn't make much difference to society anymore whether there are more or fewer farmers or whether farming is more or less efficient.

At the same time as the benefits to society have declined, the threats to society have risen. Threats to the environment, threats to the natural resource base, and threats to the quality of life of farmers, rural residents, and society as a whole all have risen. The same technologies that support our large-scale specialized system of farming are the source of these threats.

The same industrial technologies that have allowed increased agricultural productivity have now become the primary focus of growing public concerns. Commercial fertilizers and pesticides—essential elements in a specialized, industrialized agriculture—have become a primary source of growing concerns for environmental pollution. Industrialization also has made food production dependent on nonrenewable energy. Agriculture, developed for the purpose of converting solar energy to forms useful to humans, has been transformed into a system of production that uses more nonrenewable fossil energy than it produces in food energy.[5]

Historically, industrial systems have degraded the environment and depleted not only the natural resource base but also the human resource base. Henry Ford is quoted as once saying the biggest problem in running a factory is that you have to hire whole people when all you need is two hands. Large factory farms transform independent decision makers into farm workers, into people who only know how to follow instructions but not how to make decisions. The industrialization of agriculture may have made sense as long as the farmers who were displaced could find more productive employment elsewhere in the larger economy. However, the days of well-paying factory jobs are gone. American industries are reducing, not increasing, employment at all levels. Robots and computers are replacing people and eventually will do anything and everything that can be done without thinking. American industry simply doesn't need any more displaced farmers.

Rural communities most certainly have not benefited from the industrialization of agriculture. As farms have grown larger and more specialized, agriculturally dependent rural communities have withered and died. Larger farms meant fewer farms and fewer farm families to support local retail businesses, schools, churches, and public institutions. Large industrial farms typically find it more profitable to conduct their business elsewhere. The fundamental purpose of agricultural industrialization was to make it possible for fewer farmers to produce our food at lower costs. If farmers can no longer make our food cheaper and rural people have no place else to find work, why should we continue to industrialize agriculture?

There is growing evidence that in many other sectors of the

economy the process of industrialization is slowing, stopping, and even reversing. Alvin Toffler in his book *PowerShift* points out that many forecasters simply present unrelated trends as if those trends would continue indefinitely, without providing any insight into how the trends are interconnected or what forces are likely to cause them to reverse.[6] The agricultural media, both professional and popular, is filled with such forecasts for the future of agriculture.

Toffler contends that the forces of industrialization have run their course and are now reversing, the industrial model of economic progress is becoming increasingly obsolete, and the old notions of efficiency and productivity are no longer valid. He contends that mass production is no longer emblematic of the modern business operation. The new, modern model is production of customized goods and services aimed at niche markets, constant innovation, and focus on value-added products and specialized production.

He writes that "the most important economic development of our lifetime has been the rise of a new system of creating wealth, based . . . on the mind."[7] He contends that "the conventional factors of production—land, labor, raw materials, and capital—become less important as knowledge is substituted for them."[8] "Because it reduces the need for raw material, labor, time, space, and capital, knowledge becomes the central resource of the advanced economy."[9] The linear, sequential systems that characterize industrial production are being replaced with networks of simultaneous systems of production. Synergism is replacing specialization as the primary source of productivity.

Drucker, in his book *The New Realities*, talks of the "post-business society." He writes, "The biggest shift—bigger by far than the changes in politics, government or economics—is the shift to the knowledge society. The social center of gravity has shifted to the knowledge worker. All developed countries are becoming post-business, knowledge societies. Looked at one

way, this is the logical result of a long evolution in which we moved from working by the sweat of our brow and by muscle to industrial work and finally to knowledge work."[10]
Robert Reich, U.S. secretary of labor during the Clinton administration, addressed future trends in the global economy in his book *The Work of Nations*. He identifies three emerging broad categories of work in relation to competitiveness within the global economy: ~~routine production work, in-person service, and symbolic-analytic service.~~[11] Routine production workers typically work for large industrial organizations. They make their living by doing physical labor, following directions, and carrying out orders, rather than by using their minds. In-person service, like production work, entails simple and repetitive tasks but involves services that must be provided person-to-person such as retail sales workers, waiters, janitors, flight attendants, and security guards.

Symbolic analysts are the "mind workers" in Reich's classification scheme. They include all problem solvers, problem identifiers, and strategic brokers, such as scientists, engineers, bankers, doctors, lawyers, real estate developers, and consultants of all types. Like Toffler and Drucker, Reich believes that the future belongs to symbolic analysts, or mind workers, rather than routine production workers or in-person service providers.

Drucker points out another important, fundamental difference between knowledge work and industrial work: whereas industrial work is fundamentally a mechanical process, the basic principle of knowledge work is biological in nature. He concludes that this difference in organizing principles may be critically important in determining the future size and ownership structure of economic enterprises. Other things being equal, ~~the smallest effective size is best for enterprises based on information and knowledge work. There is nothing inherently to be gained by consolidating control of information and knowledge.~~

But if all this is true, why are we currently seeing the rapid industrialization in some sectors of the agricultural economy, specifically in hog and dairy production? As Joel Barker points out in his book *Paradigms: The Business of Discovering the Future*, new paradigms (including new developmental models) tend to emerge while the old paradigm, in the minds of most people, is doing quite well. Typically, "a new paradigm appears sooner than it is needed and sooner than it is wanted."[12] Consequently, the logical and rational response to a new paradigm by most people is rejection. New paradigms emerge when it becomes apparent to *some* people, not necessarily many, that the old paradigm is incapable of solving the important problems of society. Paradigms may also be applied in situations where they are not well suited, thus creating major new problems while contributing little in terms of new solutions.

American agriculture provides a prime example of overapplication of the industrial paradigm. The early gains of appropriate specialization in agriculture lifted people out of subsistence living and made the American industrial revolution possible. But the potential societal benefits from agricultural industrialization were probably largely realized by the late 1960s. The more recent "advances" in agricultural technologies may well have done more damage to the ecologic and social resource base of rural areas than any societal benefit they may have created in terms of more efficient food production.

The industrialization of agriculture probably lagged behind the rest of the economy because its biological systems were the most difficult to industrialize. Agriculture by nature doesn't fit industrialization; it had to be forced to conform. Consequently, the benefits were less, the problems are greater, it became fully industrialized last, and it likely will remain industrialized for a shorter time.

In fact, a new postindustrial paradigm for American agriculture is already emerging under the conceptual umbrella of sustainable agriculture. It has emerged to solve problems created by the industrial model, primarily pollution of our environment and degradation of our natural resource base. However, this new paradigm seems capable of creating benefits the industrial model is inherently incapable of creating, such as greater individual creativity, greater dignity of work, and more attention to issues of social equity and justice.

The sustainable agriculture paradigm is consistent with the visions of a postindustrial era of human progress shared by Toffler, Drucker, Reich, and others. Sustainable agriculture is management intensive rather than management extensive. Sustainable systems must be individualistic, site-specific, and dynamic. Thus, sustainable farming is inherently information, knowledge, and management intensive.

Complexity, interdependence, and simultaneity are fundamental elements of the sustainable model, which is clearly biological rather than mechanical in nature. For such systems, size and form must follow function. In biological systems, individual components of ecosystems must conform to their ecological niche. Big, specialized farms will be sustainable only if their ecological, social, and economic niche is equally large and homogeneous, which is not the situation on most large farms today. It will take mind work, not physical or economic muscle, for farmers of the future to find a niche where they carry out their function by means that are ecologically sound, economically viable, and socially responsible.

Why should we stop promoting the industrial paradigm of farming? Because there is growing evidence that it is obsolete, old-fashioned, out of date, and may well be doing more harm than good. Why should we stop promoting the industrialization of agriculture? Because a new postindustrial model is emerging that deserves a larger share of our time and attention.

Many of my colleagues will respond that we do not promote industrialization or any other particular model of farming. But we do. The agricultural establishment, including agricultural colleges, may not intentionally promote industrialization, but it is nonetheless promoted by their attitudes and actions.

Old, once-successful paradigms such as industrialization often collect a host of avid, but unwitting, advocates.[13] Advocates of industrialization tend to apply the industrial paradigm—unconsciously, spontaneously—to any problem that arises. They separate, sequence, analyze, and organize as a matter of standard operating procedure. Integration, simultaneity, synthesis, and spontaneity are missing from their mental problem-solving toolbox. They automatically look for gains from specialization, never synergism, regardless of the nature of the problem. In their minds, there are no logical alternatives to industrialization.

In fact, there are logical alternatives to industrialization. Success in the future will simply require new ways of thinking. In closing, I return to Peter Drucker's *Post-Capitalistic Society*: "In the knowledge society, into which we are moving, individuals are central. Knowledge is not impersonal, like money. Knowledge does not reside in a book, a databank, a software program; they contain only information. Knowledge is always embodied in a person, carried by a person; created, augmented, or improved by a person; applied by a person; taught by a person, and passed on by a person. The shift to the knowledge society therefore puts the person in the center."[14]

We need to quit promoting industrialization because it detracts from our fundamental purpose as an academic institution. That purpose is to build the productive capacity of people—to promote the public good by empowering people with the knowledge they will need to be productive in the postindustrial century of human progress. This is why we need to stop promoting the industrialization of agriculture.

‡ Presented at the Harold F. Breimyer 1995 Agricultural Policy Seminar, University of Missouri, Columbia, Missouri, November 16–17, 1995.

3

Corporate Agriculture and Family Farms

At the turn of the twentieth century, America was still an agrarian nation. In 1900 over 40 percent of the people in the United States were still farmers and well over half still lived in rural areas.[1] A hundred years later, at the turn of the twenty-first century, less than 2 percent of Americans called themselves farmers and only around 25 percent lived outside major metropolitan areas. The number of farms in the United States peaked at more than six million in the 1930s and has since dropped to less than two million. By the 1990s even those families who lived on farms relied on nonfarm income for about 90 percent of their household incomes.[2] During the twentieth century, America was transformed from an agricultural to an industrial nation.

Some scholars associate the word "industrialization" with the transformation of an economy from agriculture to manufacturing as the primary source of productivity. However, such a transformation is the natural consequence of applying an industrial model or paradigm in the development of a nation's natural resources. A fundamental characteristic of the industrial paradigm is specialization. Thus, industrialization naturally leads to some people specializing in food and fiber production, freeing others to manufacture the other things associated with an industrial economy.

In earlier times, specialization was referred to as "division of labor." Early industrialists observed that if a group of laborers who were each producing an item (i.e., transforming raw materials into finished products) would instead specialize in performing only one or two functions in the production process, they could perform each task more efficiently. By specializing

33

and working together, so that different functions were performed by different people, the group of laborers could greatly increase their collective productivity.

To facilitate such specialization, each function in the production process had to be standardized so that each specialized step in the process would fit together with the others. Specialization and standardization then allowed production processes to be routinized, and possibly mechanized, which greatly simplified the production management process. This allowed control of production to be centralized or consolidated, with fewer people making decisions but with each manager controlling the use of more land, labor, and capital. Today, we commonly refer to the economic gains from industrialization as economies of scale. Industrial development is characterized by specialization, standardization, and consolidation of control.

The transformation of American agriculture has followed the classic process of industrialization. Diversified farming operations gradually become more specialized—first specializing in livestock or crops, then in particular crops or species of livestock, and finally into specific phases of production for a specific crop or species of livestock. For example, today we have separated beef production into cow-calf, stocker cattle, and cattle feeder operations, which are separate from producing feed grains, soybeans, and hay, and from grain handlers, livestock truckers, and so on. We have separated the functions that once were performed on a single diversified farm into a number of specialized, standardized processes that are performed by separate enterprises all across the country. And in the process, we have made it not only possible but also more economically efficient to consolidate the decision making, bringing all of these specialized functions together under the control of far fewer farmers, ranchers, and other decision makers who manage far larger business enterprises.

Industrialization also results in separation and specialization with respect to the basic economic resources—land, labor, capital, and management. Some people own land, others perform labor, others provide capital, and others manage. As agricultural operations have grown larger, they have required larger amounts of capital. First, family farms were incorporated so that families could keep their capital intact as their farms were transferred from one generation to the next. But eventually, the most economical size of an operation exceeded the financial capabilities of most family corporations. Publicly held corporations are able to assemble capital from many sources, providing an almost unlimited ability to finance any economically competitive business operation. Thus, it seems inevitable that an industrial agriculture ultimately will come under the control of large publicly owned corporations. So today, American agriculture is in the final stage of industrialization—the corporatization of command and control.

In agriculture today, some people are landowners, some are agricultural workers, some own stock in agricultural corporations, and others manage agribusiness enterprises; there are relatively few real farmers left in America. Many farms, particularly the larger farms, have become corporately controlled contract production operations, and in the process, farmers have been replaced by corporate decision makers and contract farm workers. The complete corporatization of agriculture—the final stage of industrialization—would mean the end of real farming in America.

So what difference does it make whether farmers or corporations control American agriculture? Is there any real difference between farming and corporate agriculture? First, there is no useful formal definition of farming. A commonsense representation of farming in America is the traditional family farm. However, people also disagree on what constitutes a family farm. The most common definition probably is a farm in which members of the same immediate family own the land, do most of the labor, and make all the important management decisions. By one criterion or another, this definition probably would exclude most commercial farms today. It most certainly would exclude corporate contract production, in which the farmer's main contribution is borrowed capital and low-skilled labor, with the agribusiness corporation making virtually all the important management decisions.

The U.S. Department of Agriculture excludes from the definition of commercial farms those operations reporting less than $50,000, or even $100,000, in annual sales. Anything smaller is called a "hobby farm" or a "residence farm."[3] If we exclude these small "noncommercial" farms from the definition of family farms, then there are virtually no family farms left in America. Nearly all larger farms either rely on rented land or hired labor, or are contract operations.

However, I simply do not believe that a true family farm can be defined in terms of dollars of sales or percentages of land, labor, capital, or management provided by a family. A true family farm is one in which the farm and the family are inseparable parts of the same whole. If a farm is run as a separate business enterprise that simply earns income for the family, it is not a family farm. If the organization and management of the farm doesn't reflect the preferences, abilities, and aspirations of the members of the family, it is not a family farm. If a farm is not managed in a way that reflects the family's concern for their neighbors and the family's commitment to the community, it is not a family farm. Finally, if the operation of the farm doesn't reflect the ethical and moral values of the family, it is not a family farm. On the other hand, if the farm and the family are inseparable, then it's a family farm, regardless of size and regardless of who provides the factors of production and in what proportion.

A family farm can be operated on rented land and borrowed money, but the family must put much of themselves into the farm, in terms of their labor and their management, if they are truly to be a part of the farm and the farm a part of them. The economic returns from a family farm may be far more than enough to meet the needs of the family, or alternatively, the farm may show no profit at all. Family farming is not a simple matter of economics. A family farm provides recreation, education, a place to live, a place to raise a family, a place to relax and to find harmony with nature. Such things, if they could be bought, would cost thousands of dollars for an urban resident, but they all come as a natural part of a true family farm. A family farm also can help meet the social and spiritual needs of the family, regardless of whether it contributes to their economic well-being. A family farm reflects the physical abilities, the mental capacities, and the spiritual values of the family. The farm is as much a part of the family as the family is a part of the farm.

The process of industrialization has systematically destroyed family farms all across America. The sole focus of industrialization is on operational and economic efficiency. There is nothing in the industrial model to help build, or even maintain, the productive capacities of people. In fact, specialization and standardization diminish people's mental capacities because they focus on doing fewer things by the same means as everyone else, simply responding to directions or orders given by someone else. With industrialization, fewer people are given the opportunity to think—to be creative, innovative, or entrepreneurial.

There is nothing in the industrial model to help build or even maintain interpersonal relationships among people. In fact, specialization and separation virtually tears people apart, within families, within communities, and within society as a whole. People go their own way, do their own thing, and only relate to others through impersonal transactions rather than personal interaction. With industrialization, few people are given opportunities to come up with new and different *right* answers, which arise from the synergy of people thinking together.

Perhaps most importantly, nothing in the industrial model

helps to build or even maintain the ethical and moral values of individuals, families, communities, or society as a whole. In fact, once industrial operations come under corporate control, they systematically seek to destroy all social and moral constraints in the pursuit of self-interest and greed. Corporations are not people; they exist only on paper as legal, economic entities. They are created for the sole purpose of facilitating the accumulation of capital to finance large-scale industrial enterprises. Once the management of a corporation becomes effectively separated from its corporate investors, as with most publicly held corporations, the sole motives of the corporation are to make profits and grow. Most public stockholders have no commitment to or actual control over the companies in which they own shares; they invest only to earn dividends or capital gains from rising stock prices. The managers of such corporations have no choice but to maximize corporate profits and growth for the benefit of their stockholders, otherwise they will be replaced. A corporation has no heart or soul; it values only profits and growth.

The complete corporatization of American agriculture would remove the last vestiges of the American family farm. Corporatization takes the family, and even the farmer, out of the agricultural production process. Farmers are inclined to do things *their* way, whereas in corporate production, it is the corporate way or no way. And spouses or children certainly have no place in a corporate business operation. The corporate producer can't afford special considerations given to neighbors or community or to stewardship of the environment. Profit margins are too thin in corporate agriculture. Contributions to civic or charitable causes must yield economic rewards in the form of fewer social or environmental constraints to the business.

In addition, corporate agriculture must deny all ethical and moral responsibility for its actions. Much of its economic advantage comes from its willingness and ability to exploit local workers and the natural environment for corporate gains. In fact, the economic advantages of corporate agriculture would largely disappear if corporations were required to pay living wages for labor and were forced to dispose of their wastes by means that protected the natural environment and the health of their workers and neighbors. Corporate agriculture prospers by doing things that true family farmers simply would not do, because real farmers have hearts and souls. Corporate agriculture seeks to discredit and destroy the concept of family farming in order to eliminate any viable alternative to their unrestricted pursuit of ever more profit and growth.

Fortunately, a new model or paradigm for farming is emerging to address the growing deficiencies of industrial agriculture. This new paradigm is sustainable agriculture. The issue of agricultural sustainability was first raised because of an increasing realization that agricultural industrialization was destroying the natural environment. Industrial farming methods were mining the soils of nutrients, allowing soils to erode, and depleting fossil energy and other nonrenewable resources. In addition, the commercial pesticides and fertilizers essential for industrial farming were polluting groundwater and streams and were degrading the natural environment. Some people were beginning to understand that an industrial agriculture was not ecologically sustainable.

Questions of ecological sustainability led to questions of economic and social sustainability. Farmers began to realize that as industrial technologies allowed them to cultivate more land or to raise more livestock, profits per bushel or per head became narrower, leaving them no better off than before. They were farming more land, borrowing more money, hiring more laborers, and working harder, but they were earning no more for themselves than before. In addition, they realized that each round of new technology meant that some of their neighbors would have to go broke so their farms then would come up for sale. To survive, farmers had to be able to gain control of their neighbor's land. The survivors began to realize that eventually it would be their farms on the auction block. People slowly began to see that an industrial agriculture is not economically sustainable—at least not for farmers.

Questions of economic viability were followed by questions of social sustainability. Life is not just about making money, and environmental stewardship is not just about preserving nature. The economy and the environment are important because they contribute to the quality of life of people. But our quality of life also is affected by our relationships with other people—within families, communities, nations, and human society. Crime, depression, broken homes, and isolation are some common symptoms of a society whose members have lost their ability to relate to one another. Competition, confrontation, litigation, drug use, criminal activity, and war characterize societies that have depleted their social capital.

Industrialization encourages people to treat other people as interchangeable parts in machines, as nameless factors of production, as adversaries to be conquered, as something to be exploited or used up. Human relationships are reduced to contracts and transactions. There is little room for caring, sharing, or loving your neighbor in an industrial society. Some people began to realize that industrial agriculture was destroying farm families and rural communities and was contributing to the degradation of American society—that industrial agriculture was not socially sustainable.

Some people contend that the concept of a sustainable agriculture is still undefined, that they can't support it because no one really knows what it means. This quite simply is not true. People may disagree on the specific words, but there is a consensus among all who take the issue seriously that a sustainable agriculture is "an agriculture that is capable of meeting the needs of the present while leaving equal or better opportunities for the future." The concept of sustainability applies the Golden Rule across generations: we should do for those of future generations as we would have them do for us, if we were of their generation and they were of ours. We must find ways to meet our needs, all of us who are here today, without diminishing the ability of those of future generations to meet their needs as well.

A sustainable agriculture must be capable of maintaining indefinitely its productivity and usefulness to society. A sustainable agriculture must possess three fundamental characteristics. It must be ecologically sound, economically viable, and socially responsible. The three dimensions of sustainability are like the three dimensions of a box. A box without height, length, or width is quite simply not a box. Any system of farming that lacks ecological, social, or economic integrity is quite simply not sustainable. This is not a matter for debate; it is just plain common sense.

This new paradigm for agriculture is being developed by thousands of farmers across the North American continent and all around the world. These farmers are developing the replacement for the industrial model of agriculture. Farming sustainably is no simple task, but thousands of farmers are finding ways to succeed. They may carry the label of organic, low-input, alternative, or no label at all, but they are all pursuing the common economic, ecological, and social goal of sustainability. By their actions, these farmers are defining a new kind of family farming.[4]

These new farmers are diverse, but they also share much in common. They share a common pursuit of a higher and broader concept of self-interest. They are not trying to maximize profit but instead are seeking sufficient profit for a desirable quality of life. They recognize the importance of relationships, of family and community, as well as income, in determining their overall well-being. They accept the responsibilities of ethics and stewardship, not as constraints to their selfishness but as opportunities to lead successful lives.

These new farmers are succeeding by exploiting the inherent weaknesses of industrialization—of excess specialization, standardization, and consolidation of decisions. Instead, they rely on the advantages of diversity, individuality, and interdependent relationships.

The new farmers manage their land and labor resources to accommodate the diversity of nature and thus realize the benefits of nature's natural productivity.[5] By managing more intensely the land and labor resources internal to their farms, they reduce their reliance on externally purchased inputs, thus reducing costs while protecting their land from degradation and the natural environment from pollution.

The new farmers also value the natural diversity of their customers. They realize that different people naturally have different tastes and preferences, and they tailor their products to meet the unique needs and wants of their customers. They market to the growing number of people who care about where their food comes from and how it is produced. Perhaps most importantly, they cultivate personal relationships with those who care about who produced their food. They create value, increase profits, and create community through relationship marketing.

New farmers utilize their unique skills and abilities rather than blindly adopting enterprises or production methods that seem to work for others. They find markets for the things they want to produce and can produce well rather than try to produce and sell things that seem to be profitable for others. They follow their unique passions. They succeed by doing what they do best and what they like best, by valuing their individuality.

In general, the new farmers succeed by utilizing valuable resources, meeting important needs that cannot be met, or not met as well, through industrialization. Rather than seeking opportunities to specialize, standardize, and consolidation, they seek opportunities for diversification, individualization, and interdependent relationships.

Contrary to popular belief, sustainable agriculture does not mean going back to the past; it means going forward to the future. The principles of diversification, individualization, and decentralization are no older or newer than are the principles of specialization, standardization, and centralization. The fundamental challenge is identifying the principles that are more appropriate for solving the problems or realizing the opportunities of today. The industrial era is of the past, the future belongs to postindustrial knowledge-based systems. The future belongs to systems capable of empowering people to be creative, innovative, productive individuals—not just cogs in some big industrial machine. Sustainable agriculture empowers *people* to be productive.

Whether Americans will have the wisdom to develop a sustainable system of farming is a question that only time can answer. However, sustainable agriculture represents the best hope for the future of family farming, or even true farming, in America. Those who pursue a future in farming must be willing to commit their bodies and minds to farming. The parts of us that work cannot be separated from the parts of us that think. Sustainable farmers will be working thinkers and thinking workers. Those who pursue a future in farming must be willing to commit their heart and soul to farming. The part of us that loves and believes cannot be separated from the part that works and thinks. We must have the wisdom and courage to live as whole people rather than allow ourselves to become compartmentalized and isolated into economic, ecological, and ethical boxes.

Our common sense tells us that we are made up of body, mind, and soul. Our common sense tells us that the personal, interpersonal, and spiritual dimensions of our lives are inseparable and all three are important to our quality of life. Our common sense tells us that our farms must be economically viable, ecologically sound, and socially responsible—that all three are necessary and none alone is sufficient. Our common sense tells us that our quality of life is better when we care about others and when we are good stewards of nature, as we accept responsibility for taking care of ourselves. Those who want to pursue a future in farming need only rely on their common sense.

Wendell Berry, a Kentucky farmer and writer, has clearly articulated the critical nature of connections among people and between people and the land that are necessary for farming sustainably. "If agriculture is to remain productive, it must preserve the land and the fertility and ecological health of the land; the land, that is, must be used well. A further requirement, therefore, is that if the land is to be used well, the people who use it must know it well, must be highly motivated to use it well, must know how to use it well, must have time to use it well, and must be able to afford to use it well."[6]

To sustain agriculture we must have successful farmers on the land who understand the land and are committed to caring for it. The future of farming in America depends on finding ways for farmers to make a decent living while loving each other and loving the land—not on finding new industrial technologies for agriculture. The future of farming in America depends on having real people on the land whose lives are inseparable from the land. In a very real sense, the future of America is in the hands of the new family farmers.

¦ ¦ Presented at the Conference of National Block and Bridle Clubs, sponsored by the American Society of Animal Science, St. Louis, Missouri, January 20, 2001.

4

The Corporatization of America

We Americans are a fiercely independent people, right? We truly value our freedoms of speech, religion, and privacy, and our freedom to use our personal property as we see fit. We are fiercely independent about personal things. We don't want the government or anyone else telling us what we can or can't do. However, in matters that relate to our public life—to our roles in the economy, politics, and society in general—we seem more than willing to depend on others.

We let someone else decide what's in and what's out in clothes, cars, hairstyles, and soft drinks; most of us are more than willing to follow the trendsetters. We let someone else decide who will run for office and who ultimately is elected to office at the local, state, and national levels. We don't have time to waste on politics, although we can find time to complain about the stupid decisions our politicians make and the taxes we have to pay to support them. We let someone else determine the kind of society we live in, which types of behavior are socially acceptable and which are not, what's considered moral and ethical, and what's not. We leave such things to theologians and philosophers; we just aren't interested in such esoteric matters. While boldly claiming our personal independence, we depend on others to shape the economic, social, and ethical environment in which we live our lives.

We most certainly are not independent economically. We have to buy nearly everything we need from someone else and we have to work for someone else to get the money to buy those things. In economic terms, we are specialists; we do one thing for a living and depend on other specialists to provide the things we can't provide for ourselves. In addition, most of us work for some corporate business organization that makes all of our major workplace decisions for us. At work, for the most part, we do what we are told, because we think we have to in order to keep our jobs and to survive. We have no true economic independence.

We are not independent politically. We don't bother to educate ourselves on the issues. We don't participate in the process of getting people and issues on the ballots, so we don't even have a chance to vote for the people and things we want. If we participate at all, we depend on political parties, political action committees, and other special interest groups to define the issues and to articulate our political positions for us. Even when we take the time to vote, we don't vote independently. We vote for one of the two major parties, or we vote for some "independent" third party, instead of voting as individuals.

When we do exert our independence, we tend to be dominating and exploitative. We compete; we feel we must win. We must beat someone else or profit from someone else; we must use someone else for our own benefit. Without someone else to beat, we have no way to win, no way to succeed, and others have no way to win or succeed except by beating us. In reality, we are hopelessly dependent on a system that demands that we act as either victor or victim, thus encouraging us to exploit each other.

Through our lives of dependence, we have become only parts of whole people. We have let big parts of ourselves become little parts of thousands of others, and most importantly, we have become little parts of nonhuman corporate organizations and systems over which we have no influence or control. We cannot be independent because we are no longer whole people; we have lost control of our individual lives. The problem most certainly is not that we have too many personal relationships. The problem is that our relationships are dependent rather than interdependent in nature.

Steven Covey in his book *Seven Habits of Highly Effective Peo-*

ple writes about dependence, independence, and interdependence.[1] Independence may be defined as an ability to survive and thrive using one's own resources—without relationships with others. Interdependence is defined as relationships of choice rather than necessity—relationships between independent people who choose relationships that make both lives better. Dependence is defined as relationships of necessity rather than choice—relationships among people who can't survive without each other. People in a dependent relationship need to take more from the relationship than they can possibly give to it. Dependent relationships are parasitical; they are inherently exploitative and thus tend to be mutually destructive.

Dependence and exploitation are not limited to relationships among people. In America, we not only exploit each other but we also exploit our natural environment. We are parasites of nature; we extract natural resources and pollute natural ecosystems while doing little if anything to renew and restore the things we use up or destroy. We are sucking the life from each other and from the whole of creation. We are destroying the very things on which our own quality of life and the long-run survival of humanity depend. In spite of our boast of being fiercely independent, we have become totally dependent on systems of economics, politics, and ethics that are quite simply not sustainable.

The good news is that we can still break free from these destructive dependencies. However, independence is not the answer. We must move beyond *independence* to build *interdependent* relationships of choice—relationships that are mutually supportive rather than mutually exploitative. But first, we will have to become whole again. We will have to replace the broken and missing parts of ourselves and of society, but it won't be nearly as difficult as building from scratch. We don't need to become completely independent in order to choose interdependence, but we do have to stop exploiting each other. We can't become completely independent of our natural environment, but we must stop exploiting it. We need to become sufficiently independent to break free from our *unnecessary* dependencies. We must be sufficiently secure within ourselves to refuse to participate in relationships that force us to exploit or be exploited.

To break free of the grasp of destructive dependence, we need to understand the nature of the forces that hold us. Our dependence is a reflection of the society in which we live. Over the past several decades, America has evolved from a capitalist to a corporatist economy and from a democratic to a corporatist society; we have traded democratic capitalism for corporatism. And in the process, we Americans have lost our independence.

Corporatism is defined as "the organization of a society into industrial and professional corporations serving as organs of political representation and exercising some control over persons and activities within their jurisdiction."[2] Corporatism means that we participate in society not as individuals but as members of groups, which not only represent us but also exert control over us. Corporatism means that we participate in the economy, not as individuals but as members of organizations—as workers, owners, or managers of corporations. Corporatism means that we participate in the political process, not as individuals but as members of organizations—as members of labor unions, corporate business organizations, political action committees, or other special interest groups. Corporatism means that we let someone else make most of our economic and political decisions for us.

Corporatism is a natural consequence of the process of industrialization. The processes of specialization, standardization, and consolidation of control characterize the industrial paradigm. Specialization, with each person or unit performing fewer functions, allows each function or step of a production process to be performed more efficiently—that is, division of labor. Standardization allows the various specialized functions to be controlled and integrated into an efficient overall production process—that is, assembly-line production. Specialization and standardization allow, in turn, efficient centralization of management and consolidation of control—that is, economies of scale.

Economies of scale allow fewer firms or business organizations to grow larger and thus to gain greater control over the total output of an industry. As business firms become fewer and larger, they acquire increasing market power, and with this power, the ability to reduce wages and buying prices and to increase selling prices. This leads to further economies of *size*, still greater market power, and chronically declining competitiveness of markets.

Labor unions and other special interest groups have emerged to counteract the power of large industries to exploit their workers, civil society, and the natural environment, creating additional corporate organizations. However, the same organizational structure that characterizes private for-profit corporations also characterizes nonprofit special interest groups and government organizations. Each organization is made up of divisions, departments, workgroups, and so on, which perform specialized, standardized functions. Control of these organizations is centralized, allowing a few key decision makers to exert control over the people within the organization while claiming to represent them to the outside world. The corporation speaks for its stockholders and employees, the labor union speaks for its members, and political action committees speak for their contributors. People participate in society *through* these various types of corporate organizations—not directly, as independent individuals.

During my professional career, I have lived through the industrialization and corporatization of American agriculture. The motives invariably were economic. Farmers saw the opportunity to profit from adopting new agricultural technologies—new machines, fertilizers, pesticides, or business management strategies. Each new technology promised lower costs, and thus, greater profits. However, these new technologies inevitably allowed farmers to specialize, to standardize and mechanize, and to produce more than before—to farm more land, produce more per acre, manage more workers, or use more capital.

The profits inevitably went primarily to the innovators—those willing and able to take the risks of adopting unproven technologies. The early adopters realized some profits but less than the innovators, as increased production invariably caused prices to fall. The laggards eventually were forced to adopt the innovations, not to make profits but in order to survive, as prices dropped below their costs of production. However, the failure of some was necessary so that others might acquire more land so that they could reap the full benefit of the new industrial technologies. As the farms became fewer, the surviving farms became larger. The same amount of land was still farmed as before, but by larger, more industrialized farming operations.

Why should people in general be interested in what I have seen happen to farmers? This same thing has happened to nearly every other segment of the American economy. This is the same process by which the craftspeople of the past were replaced by factories, by which mom-and-pop grocery stores were replaced by supermarkets, and the small dry-goods and hardware stores were replaced by the giant discount stores.

This also is the process that ultimately brings economies and societies under corporate control, the process by which countries move from capitalism to corporatism. Incorporation allows the ownership of an organization to be separated from its management and labor. Public stock offerings allow people with large amounts of capital to own shares of companies that they do not manage or work for and allow others to work for and manage companies that they do not own. The overriding motive for public investment and ownership is to realize profits and growth in economic value. Thus, corporations purposefully seek to remove all social and ethical constraints to their pursuit of ever-greater profits and growth. Anything that is legal is considered allowable, and anything profitable is deemed desirable regardless of its social or ethical implications.

The corporatization of agriculture did not become readily apparent until the 1990s, but it should have been anticipated from the earlier industrialization of other sectors of the economy. At first corporations tended to be family corporations—a means of making capital accumulated during one generation available to the next generation within the same family. Eventually, however, family corporations began forming various types of joint ventures to gain access to still more capital. At this point in the consolidation process, existing publicly held corporations in other sectors of the economy became attracted to the newly emerging corporate sector in agriculture. Older corporations either acquired or merged with the new corporations.

As corporations become larger and larger, at some point it becomes quite difficult, if not impossible, for the remaining individually owned business enterprises to survive. The sector then is in the final stages of corporatization. And beyond some point, as the corporations grow still larger and as fewer firms control an increasing share of total output, the markets are no longer economically competitive and capitalism is transformed into corporatism.

The giant supermarket chains—such as Kroger, Safeway, and Albertsons—replaced the corner grocery store by this same process. The giant department store chains—Sears, J.C. Penney, Macy's—replaced the locally owned dry goods and housewares stores by this process. The giant building-supply chains—Lowe's, Home Depot, and Builder's Square—replaced local hardware and lumberyards by this same process. And now, still larger corporations, such as Wal-Mart with its giant "supercenters," are using this same process to displace the large supermarket, department store, and building-supply chains.

This process has been defended using the theoretical principles of competitive capitalism. "If it is a result of impersonal free markets then it must be good for society" is a common defense. However, there is no theoretical economic foundation to support the prevailing belief that a corporatist economy is capable of meeting the overall needs of society. Corporatism is not capitalism. Corporations facilitate industrialization, and thus, facilitate production of ever-increasing quantities of cheap stuff. Beyond this, there is no reason to believe that corporations will serve the needs of society as well as individuals. In fact, there is every reason to believe that corporations inevitably lead to the destruction of relationships and degradation of resources on which human society ultimately must depend.

Capitalism is based on private ownership of property by individuals. But most private property in the United States today is owned by corporations, not by individuals. Classical capitalism depends on the social values and morals of the people to constrain their pursuit of individual self-interest. Corporations have no social or moral values. The only things a corporation values are profit and growth. People have hopes and dreams for the future because they have hearts and souls. Corporations have neither. In order for capitalism to work for the good of society, for the good of people, individual people must make the economic decisions, not corporations.

Capitalism is based on competition. The conditions necessary for transforming the pursuit of self-interest into societal good are no less critical today than when Adam Smith wrote his classic book, *Wealth of Nations*. But Adam Smith's "invisible hand" of competitive markets has been mangled in the machinery of industrial corporatism.

We no longer have competitive markets, at least not in terms of having a sufficient number of buyers and sellers in each market to eliminate excessive profits and pass cost savings on to consumers. It's no longer easy to get into or out of business, as is needed to accommodate ever-changing consumer tastes and preferences. We don't have accurate information about the actual qualities of the things we buy; rather, we get disinformation by design in the form of persuasive advertising. Superficial differentiations of products abound, but there is no real variety and thus very limited real consumer choice in the marketplace.

Consumer sovereignty is a thing of the past, as advertisers now create and shape consumer demands rather than respond to them.

None of the necessary conditions for competitive capitalism exist in today's economy. The American economy is moving away from market coordination toward a corporate version of central planning. The problems of the centrally planned economies of Eastern Europe were not merely a lack of sophistication in management and planning. Central planning, by government or corporation, is fundamentally incapable of effectively coordinating a complex economy.

Capitalism is based on the principle of minimum government involvement in the economy, but the government and the economy have become inseparable. The government's primary economic function under capitalism is to maintain competition. Instead, the top priority of the U.S. government has become to promote economic growth. Corporate interests permeate every aspect of government, from the making of laws to the delivery of basic public services. It's virtually impossible to run successfully for any major public office without corporate financial backing. High-level corporate and government officials swap positions regularly as they move freely through "revolving doors" between big industry and big government. The corporations have gained so much influence in government that government not only fails to ensure competition but has become a tool for corporate exploitation of both people and resources.

We in America are in the midst of a great social experiment—one being carried out by nonhuman entities that we have created and let loose to plunder the earth. A society cannot survive in the absence of effective societal restraints to moderate the pursuit of short-run self-interest. It will exploit and eventually destroy the very things that it must have to survive—productive human and natural resources. In America, we have removed all social and moral restraints to our selfishness. We have sacrificed our independence on an altar of free markets. We, the people, are the only means left by which we can end this experiment before it is too late.

Our common sense tells us that it's time re-declare our independence. It's time for a new American Revolution. Our common sense tells us that what society needs most is not more cheap stuff. We already have more stuff than we really need. What we really need now is a greater ability to get along with other people—within families, among friends, within communities, within nations, and among people of all nations of the world. What we need now is to learn to build positive, interdependent relationships. We need to learn to build one another up rather than tear one another down. We need to take care of the earth rather than destroy it.

We need to revolt against economic and political oppression because we need to help build a better world for the future of humanity. A world with far fewer wars would be a better world. A world with less crime—fewer prisons, fewer police officers, and fewer judges—would be a better world. A world with less conflict—fewer lawsuits, fewer broken families, and fewer bankruptcies—would be a better world. All of these things are possible, but only if we break free of our destructive patterns of economic and political dependence, competition, and exploitation and start building new patterns of truly interdependent relationships.

Our common sense tells us that we need to learn to lead lives of purpose and meaning. Purpose and meaning can only come from some higher level of understanding—from some higher order of which we are but a part. We cannot gain purpose and meaning from our relationships with other people or things, no matter how strong or positive they may be. We are at the same level of organization as all of the tangible things we can see and feel; we are all part of the same whole. The meaning of our lives is not derived from our relationships with one another, but instead from the relationships of all of us to the larger whole.

We need to learn to rely on the spiritual dimension of our being for insight into the unique purpose and meaning of our lives. Through this spiritual dimension, we are rewarded when we practice stewardship—when we take care of the other living things of the earth and take care of the earth itself. Through spirituality, we are rewarded for treating those of future generations as we would like to be treated by them if we were of the future and they were of the present. A world in which people respect and take care of other living things would be a better world, accepting that plants and animals provide food for people and people must give life and sustenance to them. A world in which people care for, nurture, and restore the environment for the benefit of the future as well as the present would be a better world.

The new American Revolution must begin in the hearts and souls of the people. We need to begin by declaring our independence from the various corporate organizations that control us. Independence doesn't require that we quit our corporate jobs, but we must find the courage to refuse to do anything that exploits other people or exploits our natural environment. If we can't regain our independence in the job we now have, then we may have to seek our future employment elsewhere. We should not allow a corporation to represent us that does not respect our independence.

Independence doesn't require that we drop out of every advocacy organization to which we now pay dues. But we must find the courage and the time to oppose those organizations when we do not agree with their positions on issues, and to take an active role in shaping their policies. We cannot blindly accept the positions of leaders of any special interest group as if they were our own. We may well need to drop out of organizations that are not responsive to independent members who choose to speak for themselves.

As we reclaim our personal independence, we can begin to build interdependent relationships with other like-minded people. Relationships are important—a fundamental part of being human. But our relationships need to be empowering, not weakening or depleting. As we change ourselves, we can begin to build relationships that will change our little piece of the world.

As we regain our personal independence, we can begin to form interdependent organizations to remove our dependence on corporate organizations of all types. We can create our own jobs by joining with family members and other like-minded people to pursue ventures that are economically viable, ecologically sound, and socially responsible. We don't have to become self-sufficient. But we can develop enterprises that allow us to sell to and buy from people with whom we have mutually beneficial relationships—people we care about who care about us. We can create relationship markets. We don't have to be driven to get the highest price when we sell or the lowest price when we buy. We can insist that our trades be beneficial to both to us and to those with whom we trade.

These kinds of opportunities already exist in agriculture—through farmers' markets, community-supported agriculture groups (CSAs), community food circles, and other forms of direct marketing between farmers and their customers. These are relationship markets, where the quality of the relationships, among people and between people and the land, are at least as important as the quality of the products. A group of dedicated agrarian revolutionaries is recreating the global food system locally—one farm and one community at a time—by reconnecting people with each other and with the land.

Similar movements are under way elsewhere and can be initiated anywhere they are not already developing. All it takes is a few people who realize that change is necessary and who can find the courage to help bring it about. Similar changes can transform our nonprofit organizations and special interest groups as well. We no longer need large organizations to speak for us in the political arena. We can form far smaller groups of like-minded people. These smaller groups can form alliances with other groups on specific issues on which they agree without being tied together on issues where they do not. In these days of e-mail and the Internet, such networks of political relationships can be flexible and dynamic—interdependent rather than dependent.

As we regain independence in the workplace and in politics, we can begin to reclaim our economy and our democracy. We can wrest the political process from corporations of all types. We can force corporations to serve the public interest; we have the constitutional right to demand it. We can restore harmony and balance among the economic, social, and moral dimensions of our individual and collective social lives. We can stop the exploitation of people and of nature in America and start building a sustainable society.

America today is not unlike America of the early 1900s. John D. Rockefeller formed the first U.S. trust in 1882. He persuaded stockholders in some forty different corporations to exchange their stock for shares in the Standard Oil Company of Ohio.[3] This allowed Rockefeller to consolidate management and centralize decision making across a large segment of the petroleum industry under one board of directors, which he chaired. Rockefeller exerted market power over the petroleum industry, manipulating supplies and influencing prices and profits in ways that totally contradicted the conditions of competitive capitalism. American industrialists ever since that time have attempted to follow his lead.

In 1893 the American Sugar Refining Company and the United States Rubber Company had joined Standard Oil in the merger game. A second flurry of mergers, beginning in the early 1900s, led to the formation of such well-known companies as United States Steel, DuPont, American Can, and International Harvester. Soon large corporations not only controlled the American economy but also reached deeply into the American political process as well. Politicians and elections were routinely, often openly, bought and sold through bribes, lobbying, and corporate financing of campaigns. In many respects, the economic and political situation was not unlike that of today.

But in the early 1900s, the people rebelled. They demanded political and economic reform. Reform didn't come easily, but the people found the courage to challenge the political machines. They sent a lot of new faces to Washington to represent them. At the urging of the new president, Teddy Roosevelt, the new Congress passed a number of laws designed to help strengthen and enforce the antitrust laws already on the books.[4]

During Roosevelt's two administrations, the Justice Department brought more than forty lawsuits against the corporate trusts and won several important judgments. One judgment resulted in the breakup of the Standard Oil Company Trust. The Progressive Era in American politics continued through the Woodrow Wilson administration. Civil service eventually replaced political patronage, crippling the powerful "political machines," and primary elections were instituted to select candidates for offices instead of corporate deals in smoke-filled rooms.

The Progressives were the initial advocates of such radical ideas as election of senators by popular vote, prohibition of child labor, women's suffrage, Social Security, collective bargaining by labor, full constitutional rights for minorities, and federal curbs on monopolies. Now, once again, the country is ready for some new radical ideas.

Today, the concentration of corporate industry is far greater, and consequently, markets are far less competitive than in the early 1900s. Today's corporations are multinational, meaning they exceed the span of control of any single nation and often exceed the size of national economics. Widespread corporate alliances and joint ventures add still further to the span of control of the corporate giants. However, corporations are not more powerful than the people. People have created corporations—both business and political—and people can control corporations. We have the power, if we can find the courage to use it.

The new progressive era must begin with us, the people.

As we change ourselves, we can begin to influence others. As we influence others, we can begin to change the world around us—at least our little piece of it. As each of us changes our little piece of the world, little by little the whole of the world begins to change. This is the pattern of all great social and political movements of the past.

We can't wait for some great charismatic leader to arise. We must lead this movement ourselves; the leadership must come from the people. Certainly we need to network with others and build strong relationships, both among individuals and among groups. But we need to build interdependent relationships, not simply exchange one kind of dependency for another. We need to create a new form of democratic capitalism, based not on the independence of the past or the dependence of the present, but on interdependence—relationships of choice rather than necessity.

Idealistic? No, realistic! That's the way the world changes for the better, little by little—one person at a time. Change happens, but it is change in people that makes lasting change in society. In the words of renowned cultural anthropologist Margaret Mead, "Never doubt that a small group of thoughtful, committed citizens can change the world. Indeed, it's the only thing that ever has." We have the power to change the world, if we can find the courage to use it. We can free America from corporatism but we must each find the courage to do it, one by one.

⌐⌐ Presented at the Summer Canvassers' Conference at Ohio State University, Columbus, Ohio, sponsored by the Hudson Bay Company, Golden Valley, Minnesota, July 26–28, 2001.

PART TWO

New Hope for the Future of Farming

5

Rediscovering Agriculture and New Hope for Farming

Things are not going well in agriculture. In fact, farming is in crisis. People will continue to eat and someone will continue to produce their food, but farming, at least as we have known it, is coming to an end. As agricultural production becomes increasingly specialized and standardized, decision making is becoming centralized among a handful of large agribusiness corporations. As farms continue to become larger in size, fewer in number, and increasingly under the control of these large corporations, at some point farming is no longer farming, but instead becomes agribusiness management. Farming is associated with agriculture, not agribusiness. If farming is to survive as an occupation, we must rediscover agriculture.

So what's the difference between a farm and an agribusiness, and why does it matter? First, farmers historically have worked with nature. They attempted to tip the ecological balance to favor humans relative to other species, but they still worked *with* nature. Farmers recognized that the laws of nature must prevail over human laws. Farmers depended on unpredictable weather and worked with living systems that they could never expect to completely control. Farming always was as much a way of life as a way to make a living. A farm was a good place to raise a family and farming was a good way to be a part of a community. The benefits of farming were never solely, or even predominantly, economic in nature. Farming carried with it a set of beliefs, behaviors, and customs that distinguished it from any other occupation. It was the *culture* in agriculture that made a farm a farm and not an agribusiness.

Certainly most farmers have had times when they wished

they could control the weather and times when they longed to be more independent. If they could gain more control they could reduce risks, improve production, and make their farms more profitable. It always seemed easier to achieve the social and ethical rewards of farming than to keep pace with other occupations in terms of income and return on investment. Down deep, most farmers probably knew that if they were to succeed in achieving independence and control, they would lose some of the things they valued most about farming. But little did they realize they would lose the ability to continue being farmers.

As new technologies gave producers more control over production—commercial fertilizers, pesticides, livestock confinement, and now biotechnology—they took the physical culture out of agriculture. As new farming methods made farmers more independent—mechanization, hired labor, and financial leverage—they took the social culture out of agriculture. As humans gained control over nature, they took the spiritual culture out of farming. As farmers took the culture out of farming, they transformed agriculture into agribusiness.

As new technologies and methods succeeded in freeing farming from the constraints of nature, community, and morality, agricultural production became attractive to corporate investors. Corporations place no value on working in harmony with nature; instead, they must control nature to reduce risks and to ensure profitability and growth. Corporations place no value on relationships within families, communities, or nations; instead, they must separate people because people must compete to ensure that each produces to their full economic potential. When management becomes separated from ownership, the corporation takes on a "life" of its own. The people who choose to work for corporations may be ethical and moral people, but they are powerless to change the fundamental nature of the corporations for which they work. The only things a corporation can possibly value are profits and growth.

Crisis is chronic in agriculture, but the current crisis is different. This crisis will not simply continue the trend toward larger and fewer farms, but instead will complete the transformation of agriculture into an industry. The agribusiness corporations today seem to be using the poultry industry as their model. The poultry industry is controlled by a handful of giant corporations that control everything from the genetics for breeding stock through feeding, processing, packaging, and delivery to retail outlets. A few giant multinational corporations eventually may control each commodity sector of agriculture, giving them the ability to stabilize production at levels that maximize profits for their corporate stockholders. Consumers will become nothing more than faceless markets to be exploited and farmers little more than corporate hired hands to be used until they are used up and then replaced with machines.

With corporations firmly in control of the economic system, and seemingly in control of the political system as well, where is the hope for farming in the future? How can farming families hope to compete with the giant agribusiness firms? How can people who are committed to stewardship compete against corporations that mindlessly exploit nature? How can people who are committed to being good neighbors and responsible members of society compete with corporations that mindlessly exploit other people? The answer is that real farmers can't compete with corporate agribusiness—at least they cannot compete as bottom-line, profit-maximizing businesses. So where is the hope for the future of farming?

Hope is found in those farmers who, in the midst of crisis, are rediscovering agriculture. Vaclav Havel, writer, reformer, and former president of the Czech Republic, wrote, "Hope is not the same as joy when things are going well, or willingness to invest in enterprises that are obviously headed for early success, but rather an ability to work for something to succeed. Hope is definitely not the same thing as optimism. It's not the conviction that something will turn out well, but the certainty that something makes sense, regardless of how it turns out." He goes on to write, "It is hope, above all, that gives us strength to live and to continually try new things, even in conditions that seem hopeless. . . . Life is too precious to permit its devaluation by living pointlessly, emptily, without meaning, without love and, finally, without hope."[1]

Hope is the possibility that something good *could* happen, even when the odds are against it. Hope gives us the courage to do things that make sense simply because they are the right things to do. It's the possibility that something good could result that gives us the courage to rely on our common sense and continue to do what we feel in our hearts are the right and good things to do. Our lives are simply too precious to live without hope.

Hope for the future of farming is in agriculture, not in agribusiness. This does not mean that farmers should go back to technologies and methods of the past, although some of those may have merit for the future. Instead they must choose technologies and methods that respect the fundamental nature of farming and that keep the culture in agriculture, regardless of whether they are old or new. Certainly, farming in the future must yield an acceptable economic return to the farmer's resources—land, labor, capital, and management. But an acceptable economic return does not mean the same thing as maximum profits and growth. Farmers of the future must regain the realization that there is value in relationships among people—within families, communities, and nations. Farmers of the future must regain the realization that there is value in living an ethical and moral life—in being good stewards or caretakers of nature and of human culture. These things make sense, regardless of how they turn out, and they are the right things to do. In these things, there is hope for the future.

The most important values that arise from relationships and stewardship cannot be purchased with dollars and cents, and thus have no economic value. The industrial corporation views society and nature as constraints to profits and growth. Since corporations are not human, they cannot possibly realize the social value of human relationships or the spiritual value of human stewardship. Economics and business deal only with the personal, material self. A corporation is the ultimate "economic man"; it is driven only by the need to prosper and to perpetuate its wealth.

Farmers, farm families, and consumers, on the other hand, are real people, not created corporate entities. Real people are multidimensional. We have an individual or personal self, but we also have a social or interpersonal self, and an ethical or spiritual self. As whole people, we have three layers of self.

A part of us is embodied in our relationships with other people. This part of us does not exist separate from others, and thus is not a part of our *personal* self. Its value does not exist in individuals, but only in relationships among individuals. Its value is in such things as friendship or a sense of belonging—things that yield no individual economic rewards. Humans are social animals. By nature, we value relationships with other living beings.

Most of us say that our relationships with our spouse, our children, and our friends are the most valued aspects of our lives. Yet we allow our economy to be dominated by corporations, which have no such feelings. We continue to be driven by an economic system that places no value on relationships. Economics considers families, communities, and nations as nothing more than collections of individuals. Our society is driven by an economic system that does not make sense, regardless of how it turns out. The hope of America is in our people, not in our economic system.

Beyond the interpersonal layer is the ethical or spiritual layer of self. This dimension exists only within the context of some higher order of things. Life gains its purpose and meaning from this spiritual concept of self. The purpose or meaning of a life cannot be discerned by considering only the individual. Nothing exists only for itself. If it did, it would have no value to anything or anyone else, and thus would be irrelevant to the rest of reality. Nor can purpose and meaning be derived from our relationships with other people or things. The meaning of relationships among the parts or members of anything take on meaning only when viewed from the perspective of the whole. For example, we cannot derive the purpose of the brain from its relationship with the heart or the lungs. But rather, the purpose of each organ is discernible only in terms of its function within the whole of the human body. The body, the higher order of things, is the whole within which the organs gain their purpose and meaning.

As people, we benefit ethically and morally from our acting and living in ways that we believe to be in harmony with some higher order that transcends us and all we see or touch. A belief in a higher order of things, a sense of spirituality, is a prerequisite for realizing the ethical or moral value of our actions. In this sense, it makes no difference whether our belief in God arises from what we see in nature, or whether our respect for nature arises from our belief in God. Both are rooted in a belief in some higher order. The vast majority of all people, in all nations and cultures around the world, admit to a belief in such a higher power. Yet we continue to be driven by an economic system that gives no consideration to the spiritual dimension of self. Our current system of economics doesn't make sense, regardless of how it turns out. Our hope is that people will awaken to the spiritual dimension of self.

The hope for farmers of the future is that farmers will rediscover agri*culture*. Before the mid-1900s, farming had been about working with nature by farming in harmony with some higher, unchangeable, and uncontrollable order of things. Harmony was a means of ensuring productivity, of letting nature do more of the work. But rewards also arose directly from living and working in harmony with nature. Historically, farmers valued stewardship because they felt a moral and ethical responsibility to take care of the earth for future generations. They would care for the land even if it obviously cost them more money than they could possibly ever expect to recoup in their lifetime. They practiced stewardship because it was of value to the spiritual dimension of self—it was the right thing to do—not because of personal or individual motives. This kind of farming made sense, regardless of how it turned out.

Before the mid-1900s, farming had been about working with people—in families, communities, and nations. On a family farm, the farm and the family were inseparable parts of the same whole. The farming operations were designed to build character and self-esteem in children as they grew up. Farm work kept the family together, not just because employing the whole family improved the economic bottom line, but because building a strong family was a valued purpose for farming. Farm families valued the sharing of equipment and labor with neighbors beyond just getting the work done quicker and at a lower cost. There was value in being part of a farming community. States and nations also had strong agricultural identities. People realized that changing occupations and shifting production among regions and nations do not occur without large social and cultural costs. Historically, agriculture placed a high value on human relationships. This kind of farming made sense, regardless of how it turned out.

In reality, there is less reason to believe in the future of agribusiness than to believe in the future of agriculture. Agriculture has been around for centuries, while agribusiness is less than sixty years old. Only since the mid-1900s have we allowed the economics of individual self-interest to dominate, degrade, and ultimately destroy the ethical and social values arising from farming. Farmers have been coerced, bribed, and brainwashed into believing that the only thing that really matters, or at least the thing that matters more than anything else, is the economic bottom line. The hope for the future is that farmers will realize that their blind pursuit of profits, in fact, is the root cause of their financial failure.

Farmers have been told that they are foolish to do anything more than the minimum required by law to minimize soil erosion or protect the natural environment. Major farm and commodity organizations have worked vigorously to reduce and remove environmental restrictions on industrial farming practices in the name of maintaining economic competitiveness. Yet even under existing laws, soils have eroded at rates far faster than they can ever be regenerated. We are putting agricultural chemicals into the natural environment with little more than scientific-looking wild guesses as to whether we are doing irreparable ecological damage to natural systems. Yet farmers are told that their troubles stem from too much environmental regulation. In times past, those who purposely degraded the natural environment in their pursuit of economic gain would have been labeled by the community as ethically unfit to farm. The hope for farmers of the future is that farmers will return to the stewardship ethic of the past—that they will rediscover agriculture.

Farmers have been told that they are foolish to do anything for other people unless they expect their economic return to be greater than their individual investment. Farm programs are evaluated in terms of their economic rewards to individual farmers—not in terms of their contribution to a strong and healthy society. Government programs in general are evaluated in terms of economic impacts on consumers, agribusiness, farmers, and taxpayers. Little, if any, consideration is given to the social and ethical impacts on families, communities, states, or even nations. Farmers in the past worked together because they cared about each other as people and wanted to help each other succeed. Farmers today seem to be more concerned about getting their neighbors' land than about helping their neighbors succeed. The hope for the future is that farmers will return to valuing people over profits and to building relationships with other people—that they will rediscover agriculture.

The hope for the future is in people. People in general are beginning to question the industrial agricultural system. Consumers are becoming increasingly concerned about the quality and safety of industrially produced food and are questioning the impacts of agricultural industrialization on the natural environment. Rapid growth in consumer demand for organic foods during the 1990s gives a clear indication that the public is not buying arguments of industry advocates that high-input agriculture is both safe and necessary to ensure future food supplies. Public outcry in opposition to large-scale corporate hog operations could signal the beginning of public concern for the social as well as ecological impacts of industrial agriculture. The "big hog" issue has been a widely featured story in every mass media market available. The public is becoming aware of the true nature of industrial agriculture and they don't like what they are seeing.

Genetically modified organisms (GMOs) and the World Trade Organization (WTO) may represent the strongest one-two punch yet delivered *against* the industrialization of agriculture. Biotechnology was seen as the ultimate weapon for bringing nature to its knees. It would also be the means by which industry gained control of agriculture from genetics to the retail shelf. The WTO was industry's strategy for removing the remaining constraints to exploitation of global natural and human resources. But the people are rebelling against both. European consumers have rebelled against GMOs, and their rebellion is spreading around the world. People around the world have rebelled against the WTO—blocking an early global meeting in Seattle and continuing to harass WTO delegates as they continue their negotiations. The rebellion of ordinary people against these powerful tools of agribusiness creates hope for the future of agriculture.

However, the greatest source of hope for the future is among farmers who are seeking and finding new ways to farm. They may claim the label of organic, low-input, holistic, practical, or just plain farmer. But they are all pursuing the same basic purpose by the same set of principles, trying to build farming systems that are ecologically sound, economically viable, and socially responsible.[2]

They realize that quality of life is a product of harmony among the economic, social, and spiritual dimensions of their lives. They are creating farming systems that will meet their needs while leaving equal or better opportunities for others, both today and in the future. They refuse to exploit other people or exploit the natural environment for short-run personal gain. They are building an agriculture that is sustainable over the long run, not just profitable for today. They are rejecting the exploitation and competition of agribusiness; they are rediscovering agriculture.

These new farmers are the hope for the future of agriculture. Hope gives them the strength to continue trying new things, even though they are working against seemingly insurmountable odds. They suffer frustrations, hardships, and even failures; such is the nature of being pioneers. Success has not come easily, but for many it has come. And while still a small minority, the ranks of new sustainable farmers are growing across the continent and around the world. All across North America, the number of "sustainable agriculture" conferences, and the number of farmers attending each conference, continues to grow each year. Even more importantly, in the successes of even a few there is hope for the many.[3] These farmers are farming in ways that make sense, regardless of how it all turns out, and in this, there is hope.

These new American farmers are a diverse lot—young and old, female and male, families and singles, experienced farmers and new farmers. They represent wide ranges in formal education, income levels, and ethnicity. But they all share a common vision of hope in a common belief in the possibility of building better lives for themselves, for their families, and for society through sustainable farming. They are hopeful, if not optimistic, about the future of their kind of farming and their way of life. These hopeful people are the hope for the future of farming. They are rediscovering agriculture.

‡‡ Presented at the Agriculture and Rural Life Conference, sponsored by Catholic Rural Life, Bruno, Saskatchewan, Canada, March 25–26, 2000.

6

Farming in Harmony with Nature and Society

Much of human history has been written in terms of an ongoing struggle of "man against nature." The forces of nature—wild beasts, floods, pestilence, and disease—have been cast in the role of the enemy of humankind. To survive and prosper, we must conquer nature—kill the wild beasts, build dams to stop floods, find medicines to eliminate disease, and use chemicals to destroy pests. Humans have been locked in a life-and-death struggle against nature. We've been winning battle after battle, but we've been losing the war.

We humans have killed so many "wild beasts" that nonhuman species are becoming extinct at a rate unprecedented, except for events in prehistoric times now labeled as global catastrophes. Humans cannot survive, nor might we choose to survive, as the only living species on earth. How many more species can we destroy before we lose more than we can live without? How many more battles with nature can we afford to win?

We have dammed so many streams that the sediment that once replenished the topsoil of fertile farmland through periodic flooding now fills our reservoirs and lakes instead. Populations of fish and wildlife that once filled or drank from free-flowing streams, and once fed the people of the land, have dwindled in numbers or completely disappeared. Floods may occur less often now, but when nature really flexes its muscles, as in the Midwest in 1993 and 1996, nothing on earth can control the floods. How many more streams can we afford to dam before we have dammed too many? How many more battles with nature can we afford to win?

We have wiped out plague after plague that has threatened

humankind, and we now lead longer, presumably healthier lives than ever before. But new, more sophisticated diseases always seem to come on the scene as soon as the old ones are brought under control — HIV/AIDS being only the latest in a long line of great human killers. We may live longer, but that doesn't necessarily mean we are healthier, as attested to by skyrocketing medical costs of older Americans. Much of the medicine we take today is prescribed to treat unanticipated side effects of the medicines we were already taking. On average, we Americans spend more money for health care than we spend for food. How long can our new cures keep ahead of new diseases while our new medicines keep creating new "diseases"? How many more medical miracles can we afford? How many more battles with nature can we afford to win?

We can now kill most insects, diseases, weeds, and parasites using modern chemical pesticides. This has allowed us to abandon diversified family farms and to realize the lower food prices associated with a specialized, mechanized, standardized, industrialized agriculture. But we still lose about the same percentage of our crops to pests as we did in earlier times. In addition, health concerns about pesticide residues in our food supplies and in our drinking water are on the rise. Rural communities also have withered and died as industrial agriculture has displaced the farm families that once supported local schools, churches, civic organizations, and businesses. How many more pests can we afford to kill before we kill our rural communities, or even more importantly, before we kill ourselves? How many more such battles can we afford to win?

Each time we think we are winning the war, nature fights back. Nature always seems ready with a counterattack after each battle we win. Many people are beginning to lose faith in humankind's ability to conquer nature. They are concerned about whether we can win the battle with the next flood, the next disease, or the next pest that we have created in our efforts to control the last one. People are concerned with their own safety, health, and physical well-being, but they are concerned also about the sustainability of their communities and of a human civilization that continues to live in conflict with nature. They fear we cannot win our war against nature, because we are a part of the nature we are trying to destroy. They are searching for ways to live in harmony with nature—to sustain the nature of which we are a part.

A new paradigm or model for working and living in harmony with nature is arising under the conceptual umbrella of sustainability. Sustainable systems must be capable of meeting the needs of the present generation without compromising the ability of future generations to meet their needs as well. In simple terms, sustainability means applying the Golden Rule across generations—doing for those of future generations as we would have them do for us, if we were of their generation and they of ours. It's about short-run self-interest—that is, meeting our individual, present needs—but it's also about long-run common interest—leaving equal or better opportunities for others, both of the present and of the future. Sustainability requires that we live in harmony with others of the present and of the future.

The sustainable agriculture movement is but one small part of a far larger movement that is transforming the whole of human society. But a society that cannot feed itself quite simply cannot sustain itself. Human civilization is moving through a great transformation from the technology-based industrial era of the past to a knowledge-based *sustainable* era of the future, and sustainable agriculture is an essential part of that transition.

The industrial model of development is rooted in the historical premise that the welfare of people is in conflict with the welfare of nature. It assumes that people must harvest, mine, and otherwise exploit nature to create more goods and services for human consumption. Human productivity is defined in terms of one's ability to produce goods and services that will be valued, bought, and consumed by others. Quality of life is viewed as related directly to wealth and the consumption it affords, as something we might buy at McDonald's, Wal-Mart, or Disney World. The more we produce, the more we earn, and thus, the more we can consume and the higher our quality of life. The industrial model assumes that the more we can take from nature, the higher will be our quality of life.

On the other hand, the sustainable model is based on the assumption that people are multidimensional; we are physical, mental, and spiritual beings. We have a mind and soul as well as a body. All three determine the quality of our life—what we think and what we feel as well as what we consume. A life lacking physically, mentally, or spiritually is not a life of quality. The industrial model has focused on the physical body, the self—on getting more and more to consume. The sustainable model focuses instead on finding balance and harmony among all three—the physical, mental, *and* spiritual.

Spirituality is not synonymous with religion. Spirituality refers to a felt need to be in harmony with some higher unseen and unalterable order of things—paraphrasing William James, a noted religious philosopher.[1] Religion at its best is simply one means of expressing one's spirituality. Harmony cannot be achieved by changing the order of things to suit our preferences; the order is unchangeable. Harmony comes only from changing our actions instead to conform to the higher order. A life of peace and happiness is the result.

An agriculture that uses up or degrades its natural resource base or pollutes the natural environment eventually will lose its ability to produce. It's not sustainable. An agriculture that isn't profitable, at least over time, will not allow its farmers to survive economically. It's not sustainable. An agriculture that fails to meet the needs of society, as producers and citizens as well as consumers, will not be sustained by society. It's not sustainable. A sustainable agriculture must be all three—ecologically sound, economically viable, and socially responsible. And the three must be in harmony.

Some see sustainability as only an environmental issue. They are wrong. It *is* an environmental issue, but it also is much more. Any system of production that attempts to conquer nature will create conflicts with nature, will degrade its environment, and will risk its long-run sustainability. Nature provides the only means of capturing and converting significant quantities of solar energy for use by living plants and animals. Living things cannot survive on electricity generated by wind, water, or voltaic panels. By nature, organisms come to life, grow to maturity, reproduce, and die, their bodies to be consumed by other living organisms or recycled to support a future generation of life. Agriculture attempts to tip this ecological balance in favor of humans relative to other species. But if we attempt to tip the balance too far too fast, we destroy the integrity of the natural system of which we are a part. A sustainable agriculture must be in harmony with nature.

A sustainable agriculture also must function in harmony with people, since people are a part of nature, with a basic nature of their own. A socially sustainable agriculture must provide an adequate supply of food and fiber at a reasonable cost. Any system of agriculture that fails this test is not sustainable, no matter how ecologically sound it may be. But "man does not live by bread alone," and a socially responsible agriculture must contribute to a positive quality of life in other respects as well. A sustainable agriculture must meet the food and fiber needs of people, but in the process it cannot degrade or destroy opportunities for people to lead successful, productive lives. A sustainable agriculture must be in harmony with our nature of being human.

Finally, a sustainable agriculture must be in harmony with the human economy. The greatest challenge to sustainable farming is in finding ways to make ecologically and socially responsible systems economically viable as well. Our current economy favors systems that exploit the natural and human environment for short-run gains. Those who choose to protect the natural environment must forgo any economic opportunity that might result from resource exploitation. Those who show concern for the well-being of other people—workers, customers, or neighbors—must forgo any economic opportunity that might result from their exploitation. So it might seem that sustainability requires that one must sacrifice economic well-being to achieve ecological and social sustainability.

The relationships among environment, social, and economic well-being typically are treated as trade offs; people assume that we can have more of one only by sacrificing some of the others. For anyone to gain more of something, a like amount must be sacrificed by someone else. There is only some fixed quantity to be allocated among competing ends. However, this highly materialistic worldview ignores the fact that we can gain satisfaction for ourselves right now by doing things for others and by saving things for future generations. Our satisfaction is not dependent on realizing the expectations of some future personal rewards; the reward is embodied in the current action rather than the future outcome. There is inherent value in doing the right thing, in living and working in harmony. Getting more of one thing without having more of the others only creates imbalance and disharmony—making us worse off rather than better.

However, the necessity for economic viability is a very real concern, even for those who pursue harmony rather than material wealth. If our endeavors are not economically viable, we lose the right to pursue those endeavors. But how can a person make a living farming without degrading either the natural environment or the surrounding community? The standard of performance of industrial farming is dollar-and-cent costs of production, and thus industrial farming exploits its natural and human resource base to keep those costs to a minimum. So how can a sustainable farmer compete? The answer for sustainable farmers is not to compete with industrial farming but to do something fundamentally different.

This something different includes letting nature do more of the work of production—working with nature rather than against it. Industrial systems require uniformity and consistency, but nature is inherently diverse and dynamic. Harmony comes from matching what is produced and how it is produced with the unique ecological niche within which it is produced. The greater the harmony, the more of the work nature will be willing to do and the more productive the farm will be.

Finding harmony means providing people with what they need and want rather than coercing or bribing them to take what you have for sale. Industrial systems of mass production and mass distribution gain their cost advantage by treating people as if they were all pretty much the same. Harmony comes from being sensitive to the individual needs and wants of inherently diverse individuals—producing in harmony with human nature. The greater the harmony, the more valuable the product will be.

Finding harmony means reconnecting with people—as fellow human beings rather than as consumers, producers, or some other generic economic entity. Joel Salatin, a Virginia farmer and "agripreneur," refers to this as "relationship marketing."[2] When you have a relationship with your customers, they do not simply represent a market to be exploited to make a few more dollars. They are friends and neighbors that you care about and don't want to lose. When your customers have a relationship with you, you are not just another supplier to be haggled down to the lowest possible price to save a few dollars. You are someone they care about and don't want to lose. When you know, care about, and have affection for each other, you have a relationship that creates value beyond market value. You are contributing directly to each other's quality of life. You are creating a harmony that arises only among people who love one another.

Kentucky writer and farmer Wendell Berry in his book *What Are People For?* puts it more succinctly: "Farming by the measure of nature, which is to say the nature of the particular place, means that farmers must tend farms that they know and love, farms small enough to know and love, using tools and methods that they know and love, in the company of neighbors they know and love."³ Neither land nor people can be sustaining or sustained unless they are given the attention, care, and affection—the love—they need to survive, thrive, and prosper. The necessary attention, care, affection, and love come only from lives lived in harmony—among people and between people and nature.

Finally, as more farmers and customers, sharing common concerns for ecological and social sustainability, develop personal relationships in the marketplace, economic communities of interest will expand as well. Customers will be willing to pay more and farmers will be willing to provide more because they are both getting more from the relationship than is reflected in the market price. Those who might attempt to exploit these new economic communities for short-run gains—those motivated by economic value rather than ethical or moral values—are destined to find disappointment. Those who join in seeking balance among the economic, ecological, and social dimensions of their lives—among the physical, mental, and spiritual—will be rewarded. They are leading the way into a new era of sustainability in which people strive to live in harmony with each other as well as in harmony with nature.

⌐⌐ Presented at Agri-Expo '99, sponsored by the University of Missouri, Columbia, Missouri, March 23, 1999.

7

Reclaiming the Sacred in Food and Farming

Farming is fundamentally biological. The essence of agriculture is the living process of photosynthesis—the collection, conversion, and storage of solar energy. All living things are sustained by other living things. If life is sacred, then the food and farms that sustain life must be sacred as well. In fact, throughout nearly all of human history, both food and farming *were* considered sacred. Farmers prayed for rain, for protection from pestilence, and for bountiful harvests. People gave thanks to God for their "daily bread," as well as for harvests at annual times of thanksgiving. For some, farming and food are still sacred. But for many others, farming has become just another business and food just something else to buy. Those who still treat food and farming as something sacred tend to be labeled as old-fashioned, strange, radical, or naive.

However, the time to reclaim the sacred in food and farming may be at hand. The trends that have desacralized food and farming may have run, even overrun, their course. Today, there is growing skepticism concerning the claim that more *stuff*, be it larger houses, fancier cars, more clothes, or more food, will make us more happy or more satisfied with life. There is growing evidence that when we took out the sacred we took out the substance and left our lives shallow and empty. The old question "How can I get more stuff?" is being replaced with a new question, "How can I find a better life?"

The answer to the new question, at least in part, is that we must reclaim the sacred in our lives. But how can we reclaim the sacred, and how will it change the ways we farm and live? These questions will be addressed, but first we need to understand why we took out the sacred in the first place and why we now need to put it back in.

Until some four hundred years ago, nearly everything in life was considered spiritual or sacred. Religious scholars were the primary source of learning and knowledge in the so-called civilized world of that time. Kings, chiefs, clan leaders—the people who others looked to for wisdom—were assumed to have special divine or spiritual powers. It was only during the seventeenth century that the spiritual nature of the world became seriously challenged. Among the most notable challengers was the Frenchman Rene Descartes, who proposed the "spirit versus matter" dualism. "The Cartesian division allowed scientists to treat matter as nonliving, or dead, and completely separate from themselves, to see the material world as a multitude of different nonliving objects assembled into a huge machine."[1] Sir Isaac Newton also held this mechanistic view of the universe and shaped it into the foundation for classical physics.

Over time, the mechanistic model was expanded to include the living as well as the nonliving. Plants, animals, and even people were increasingly viewed as complex mechanisms with many interrelated yet separable parts, in spite of the emergence of quantum physics, which challenges the old mechanistic worldview. Reductionism, which attempts to explain all biological processes as purely chemical and mechanical processes, now dominates the applied biological sciences from agriculture to medicine.

The spiritual realm, to the extent it is considered at all, is assumed to be in the fundamental nature of things, in the unchanging relationships that scientists seek to discover. There is no active spiritual aspect of life in science, only the passive possibility that spirituality was somehow involved in the initial creation of the universe we are now exploring. The more we understood about the working of the universe, the less we needed to understand about the nature of God. The more we knew, the less we needed to believe. As we expanded the realm of the factual, we reduced the realm of the spiritual until it became trivial, at least in matters of science.

This shift in scientific thinking has been a shift from a science of understanding to a science of manipulation.[2] Over time, the goal of science shifted from increasing wisdom to increasing power. We wanted not only to understand the universe but also to dominate it. The purpose of science became to enhance our ability to influence, direct, and control. Farming was one of the last strongholds for the sacred in the scientific world. Mechanical processes, using machines to manufacture things from inanimate matter, were relatively easy to understand, control, and manipulate. Biological processes, which involve living organisms (including people), proved much more difficult to both understand and control. Farming, being a biological process involving people, proved especially difficult to manipulate and control. Farmers continued to pray for rain and people continued to give thanks for their food, long after most scientists advised us that both were either unnecessary or futile.

Science eventually succeeded in taking the sacred out of farming, at least out of modern, scientific farming. Machines took laborers out of the fields, making farming more manageable. Selective breeding brought genetic vagaries more or less under control. Commercial fertilizers gave farmers the power to cope with the uncertainties of organic-based nutrient cycling. Commercial pesticides provided simple scientific means of managing predators, parasites, and pests. Deep-well irrigation reduced the grower's dependence on rainfall. Processing, storage, and transportation—all mechanical processes—removed many of the previous biological constraints associated with form, time, and place of production. Farms became factories without roofs. Supermarkets and restaurants are just the final stages in a long and complex food assembly line. Why pray for rain when we can drill a deep well and irrigate? Why thank God for food created by ConAgra? Who needs God when we have modern science and technology?

But today, as in the seventeenth century, we are in a time of great transition. "We are at that very point in time when a 400-year-old age is dying and another is struggling to be born—a shifting of culture, science, society, and institutions enormously greater than the world has ever experienced. Ahead, the possibility of the regeneration of individuality, liberty, community, and ethics such as the world has never known, and a harmony with nature, with one another, and with the divine intelligence such as the world has never dreamed."[3] These aren't the words of a philosopher or a cleric. These are the words of Dee Hock, founder of Visa Corporation and creator of the Chaordic model of business organization.

Hock is certainly not alone in this thinking. A whole host of futurists, including Alvin Toffler, Vaclav Havel, Tom Peters, Peter Drucker, John Naisbitt, Robert Reich, and others, agree that we are living in a time of fundamental change.[4] They talk and write of a shift from the mechanistic worldview of the industrial era, where power is derived from control of capital and the technical means of production, to a new postindustrial worldview, where human progress is derived from knowledge, the new source of wealth and human satisfaction. They agree that knowledge is fundamentally biological rather than mechanical in nature, and therefore, management of knowledge will require a new science of understanding to replace the old science of manipulation.

The transition to a more sustainable agriculture is but one small part of the great transition that is taking place all across society. The questioning that is driving the sustainable agriculture issue exemplifies the broader questioning of society that is fueling the great transition. We are questioning the sustainability of agriculture because we have come to understand that our natural resource base is finite, that we and the other elements of our natural and social environment are all interconnected, that there is a higher, unseen order of things to which we must conform. Concerns for sustainability seem foolish to those who believe that human ingenuity is infinitely substitutable for natural resources, that we are separable from our environment, and that the laws of nature are merely temporary obstacles to be overcome through better science. Conflicts regarding the legitimacy of the sustainability issue are conflicts of beliefs, not of facts. There is a growing body of evidence, however, to support the legitimacy of questioning whether agriculture or any other aspect of our modern industrial society is sustainable.

In agriculture, the litany of sustainability concerns has become a familiar theme. Modern, scientific agriculture uses more fossil energy, particularly oil and natural gas, than it produces in terms of food energy. Water and air pollution—associated with commercial fertilizers and pesticides and large-scale confinement animal feeding operations—have become major public health concerns. Declining numbers of family farms—a consequence of agricultural industrialization—have left many rural communities in decline and decay, as places without a purpose.

Farmers' historical ethical and moral commitments to stewardship and community seem to have given way to concern for the economic bottom line. Gains in agricultural productivity have become more illusionary than real as the quest for market power has replaced the quest for production efficiency. There is little farming left in food production, as the farmer's role has declined and the roles of input and marketing firms have risen. Small farms are now considered largely irrelevant to agriculture, even though most U.S. farm families still live on small farms. Today, a growing sense of disillusionment and hopelessness prevails, even among larger farmers, as multinational corporations take over a larger and larger share of agricultural production through comprehensive contractual arrangements.

Similar concerns are apparent in the larger society. As population and per capita consumption continue to increase, the ultimate scarcity and depletion of natural resources—including land and fossil fuels—seem obvious to a great many people.

The environmental movement, born only in the early 1960s with Rachel Carson's *Silent Spring*, has grown to permeate global society as evidence of environmental pollution abounds.[5] The disintegration of families and communities in the relentless pursuit of wealth is beginning to have major negative impacts on our societal quality of life. Increasing drug use, violence, and crime are attributed to the decline in ethical and moral values of a disconnected society. Declining productivity of labor, a symptom of treating people as if they were machines, has led to growing underemployment and economic and social inequities. These and other factors contribute to a growing disillusionment and sense of hopelessness that permeates much of society. At a world conference of intellectuals reported in the book *Reinventing the Future*, degradation of the environment, breakdown of public and private morality, and growing social inequities between countries of the northern and southern hemispheres were identified as three of the four most critical items on the global agenda.[6]

What do these concerns for sustainability have to do with spirituality? They share a common source in the removal of spirituality from science and society. The science of manipulation, the quest for power and control, provided the conceptual foundation for the industrial revolution. The fundamental concepts of industrialization—specialization, mechanization, routinization, and consolidation of control—are based on a mechanistic worldview. The science of Descartes and Newton became a science that sought to separate, sequence, and compartmentalize processes and people. Growing concerns for ecological, social, and economic sustainability all are consequences of this industrial way of thinking. And in the mechanistic worldview supporting industrialization, there is no active role for the social or the sacred.

The science of manipulation is a science that separates—mind from matter, people from nature, people from each other, body from mind, and mind from soul. The science of modern neoclassical economics assumes the greatest good arises spontaneously from the greatest greed—the interest of society is served best by the vigorous pursuit of self-interest. The same science that made the industrial era possible is the science that eventually removed the sacred from matters of economics and politics, which in turn is removing spirituality from the day-to-day matters of both individuals and communities. We were led to believe that so-called good science would bring about success and happiness without any help from "on high," but we were misled.

Biological and social phenomena have never really fit the mechanistic, manipulative view of the world. Living things of nature had to be bent, twisted, bribed, and coerced to bring them under control. But nature inevitably fights back, and "nature always bats last." Questions of sustainability invariably can be traced to unintended consequences of treating living things as if they were inanimate, programmable, controllable machines. A science of understanding—of wisdom rather than power and control—must provide the foundation for a sustainable society.

Using almost anyone's definition, concerns for sustainability imply concerns for intergenerational equity—a need to meet the needs of our current generation while leaving equal or better opportunities for those of generations to follow. The three cornerstones of sustainable agriculture—ecological soundness, economic viability, and social equity—rest upon a foundation of intergenerational equity. Intergenerational equity, in turn, has its foundation in human spirituality. Concern for sustainability reflects a felt need for fair and equitable treatment of those of future generations, with whom we share no interests in any sense other than spiritual.

Neoclassical economic theory deals with short-run self-interest. Economic efficiency defines the optimum means of using things up. There is nothing in economics to ensure investments in renewal and regeneration necessary for long-run sustainability. Economics is about *me, now*. Conventional public choice theory deals with collective decisions concerning matters of current shared interest. There is nothing in this theory concerning allocating societal goods and services to ensure the long-run sustainability of society. Public choice is about *us, now*. Likewise, many of the current environmental concerns are related to a desire to protect *us, now*, rather than our concern for future generations. But sustainability must include concerns *for us and for them, forever*. Only the spiritual is capable of transcending the present to address the fundamental issues of long-run sustainability. Only the spiritual transcends me, us, and them, both for now and forever.

What is this thing called spirituality? First, spirituality is not religion, at least not as it is used here. Religion is simply one of many possible means of expressing one's spirituality. William James, a religious philosopher, defined the essence of religion as "a belief that there is an unseen order, and that our supreme good lies in harmoniously adjusting ourselves thereto."[7] This description embraces a wide range of cultural beliefs and philosophies, as well as religions. Thus, the more general concept of spirituality might be defined as a felt need to live in harmony with a higher, unseen order of things.

A Native American, Chief Sealth, or Seattle, is paraphrased as saying, "Whatever befalls the earth befalls the sons and daughters of the earth. We did not weave the web of life; we are merely a strand in it. Whatever we do to the web, we do to ourselves."[8]

From another culture, "the most important characteristic of the Eastern worldview—one could almost say the essence of it—is the awareness of unity and mutual interrelation of all things and events, the experience of all phenomena in the world as manifestations of a basic oneness."[9]

An example of a Polynesian worldview: "The Kahuna told me, if you are looking for God, look out at the sea. Look to the horizon. Get in your canoe and go to the horizon. When you get there, you will meet God. God is nature. God is everything."[10]

From a Jewish prayer: "And God saw everything he had made and found it very good. And he said, 'This is a beautiful world I have given you. Take good care of it; do not ruin it. . . . I place it in your hands: hold it in trust.'"[11]

Finally, from Ecclesiastes 3:1–8 in the Bible: "To everything there is a season, a time for every purpose under the sun; a time to be born a time to die; a time to plant and a time to pluck up that which is planted; a time to kill, a time to heal; a time to weep a time to laugh; . . . a time to love and a time to hate; a time for war and a time for peace."

A common thread of all these expressions of spirituality is the existence of an unseen order or interconnected web that defines the oneness of all things within a unified whole. We as people are a part of this whole. We may attempt to understand it and even influence it, but we did not create nor can we control it. Thus, we must seek peace through harmony within this order rather than attempt to change it. This harmony may be defined as "doing the right things." By doing the right things for ourselves, for others around us, and for those of future generations, we create harmony and find inner peace and happiness.

The sustainable agriculture issue ultimately is rooted in a belief in a higher order of things and that "our supreme good lies in harmoniously adjusting ourselves thereto"—in spirituality. Finding harmony with a higher order requires an understanding of that order—in wisdom, not power and control—so we may nurture nature rather than attempt to dominate or manipulate it. Sustainable agriculture means fitting farming to the farmer and the farm—not forcing either to fit some predefined prescription for progress. Sustainable farming means farming in harmony among people—within families, communities, and societies. Sustainable farming means farming in harmony with humanity—being good stewards of the earth's finite resources. A life of quality is a shared life and a spiritual life.

Quality of life is not something we can buy at Wal-Mart or Disney World with the money we earn by working or farming for the economic bottom line. Quality of life is determined by our ability to do the right things, for me, for us, and for them. Quality of life, inherently and inseparably, is personal, interpersonal, and spiritual in nature. Sustainability is about sustaining a desirable quality of life, thus sustainable agriculture is personal, interpersonal, and spiritual.

Protecting our own environment is not enough. We must conserve and protect resources for those of the future. Profits are necessary for sustainability, but profits alone are not sufficient to ensure sustainability. The pursuit of economic short-run self-interest will not ensure that anything will be left for future generations. A society without justice is not sustainable—no matter how profitable and environmentally sound it may seem. The economic, ecological, and social dimensions of sustainability are all rooted in spirituality. If we are serious about the pursuit of a sustainable agriculture, we must begin by proclaiming, up front and without compromise, the spiritual nature of sustainability.

As we reclaim the sacred in food and farming, it will change the way we farm and live. Our common sense tells us that we must have balance in our lives among the personal, interpersonal, and spiritual. Yet we are bombarded from every corner with the message that having more stuff will make us happy, that success means having more money. Or we may be told that happiness is found only in love of family and friends, and that money doesn't matter. On Sunday, the message is likely to be that happiness comes only from the love of God, that we should deny ourselves and follow Him. The basic thesis of sustainability is that all these things matter in our quest to live in harmony within the higher, unchanging order.

Reclaiming the sacred does not mean that our rewards must be delayed until the afterlife, any more than sustainability means we must sacrifice quality of life today for the benefit of some future generation. We live only in the present, not the past or the future. If we are unhappy today, achieving some future tangible goal is likely to leave us still unfulfilled and unhappy. If we are happy today, then we are quite likely to find some way to be happy in the future, regardless of whether we achieve some goal we have in mind today. The focus of faith and hope may be on things expected or hoped for in the future, but the true benefits of both are in the here and now. Faith and hope are about *here and now*, not *there and then*. Faith and hope are fruits of the spirit.

Likewise, the spiritual dimension of sustainable farming is about here and now, not there and when. Those rewards come from having adequate, not maximum, income, from having positive relationships with others, and from being a responsible steward of resources for the future. All those things have rewards for us here and now, as well as for someone else, somewhere else, at some time in the future. The reward comes from knowing that we are living in harmony with the unseen order.

A desire to reclaim the spiritual does not guarantee peace and happiness. We still must seek to understand so we may learn to accommodate, rather than dominate, and to nurture, rather than conquer. We need to be wise, not smart. We need to learn to be humble, not powerful. We need to seek and accept the spiritual in everything we see and do. We need to learn to dance with life rather than try to push life around.

To farm and live sustainably is to farm and live spiritually. Sustainability is not a religion, but it is undeniably spiritual. To farm and live sustainably, we must be willing to reclaim the sacred in food and farming.

¦ ¦ Presented at the Twelfth Annual Sustainable Agriculture Conference of the Carolina Farm Stewardship Association, Flat Rock, North Carolina, November 1997.

PART THREE

Principles of Sustainable Agriculture

8

Do We Really Need to Define Sustainable Agriculture?

A lot of time and intellectual energy has been spent attempting to define sustainable agriculture. A consensus seems to be emerging in the movement that we need to spend less time trying to define it and more time working to achieve it. But how can we work to achieve something without first defining it? We can if we can simply agree that the basic goal is agricultural sustainability, with the words "agricultural" and "sustainability" both defined in their generic sense. In fact, most of our definitional disagreements seem to stem from differing opinions concerning the means by which agriculture can or should be made sustainable, rather than the end results toward which those means are directed.

"Sustainability is a question rather than an answer," as the late Robert Rodale was fond of saying. It is a direction rather than a destination, like a star that guides the ships at sea but remains forever beyond the horizon. The question of sustainability can be asked of any ongoing activity or process, including conventional agriculture and any proposed alternative. Asking the question need not, and should not, presuppose the answer.

Reaching agreement on the goal may not be simple, but it should be achievable. First, we must agree on what is to be sustained, for whom, and for how long. If we can agree on the answers to these questions, we should be able to move toward a common goal. I believe most advocates are working to sustain agriculture for the benefit of humanity, forever (or at least as long as the sun continues to shine). Agriculture, by its very nature, is an effort to shift the ecological balance so as to favor humans relative to other species. Thus, if we sustain agriculture we are sustaining it for the ultimate benefit of humankind. I believe there is a consensus also that we are concerned about the well-being of people of this generation and for all generations to follow. I have seen no inclination to place a time horizon on how long agriculture should be sustained.

Lacking a specific time horizon, we cannot prove through empirical studies that one approach to agriculture is sustainable or that another is not. It would quite literally take forever to collect the data for such a study. Thus, we must rely instead on logic. A sustainable agriculture logically must be ecologically sound, economically viable, and socially responsible. Furthermore, these three dimensions, insofar as they relate to sustainability, are inseparable. All three are essential, and thus are all equally critical.

Most people who are concerned about sustainability recognize an interconnectedness of humanity with the other biophysical elements of our natural environment. Through agriculture, we may tip the ecological balance in our favor. But if we attempt to tip it too far or too fast, we will destroy the integrity of the natural ecosystem of which both we and agriculture are part. If we degrade our natural resources and poison our natural environment, we will degrade the productivity of agriculture and ultimately destroy human life on earth. Nearly everyone seems to agree that a sustainable agriculture must be ecologically sound.

There may be less agreement regarding economically viable and socially responsible. The social sciences of economics and sociology are fundamentally different from the physical sciences of agriculture and the natural science of ecology. Agriculture by its nature involves conscious attempts by humans to change or manage natural ecosystems. Humans are unique among species in that we make purposeful, deliberate decisions that can either enhance or degrade the health of the ecosystems of which we are a part. Thus, any question of sustainability must take into account the purposeful, conscious nature of individual and collective human actions, which are driven by the economic and social motives of people.

Sustainable systems must be economically viable, either by nature or through human intervention. In many cases, farmers have economic incentives to adopt ecologically sound systems of farming. A healthy agro-ecosystem tends to be a productive and profitable agro-ecosystem. However, inherent conflicts exist between the short-run interests of individuals and the long-run interests of society and humanity. In such cases, society must make it economically feasible, through public policy, for individuals to act in ways consistent with long-run societal interests.

Human nature is a part of nature. Even when our physical survival is ensured and our basic needs are met, humans by nature act in our own economic self-interest. We need not seek to maximize profit or wealth, but people cannot persist in actions that are inconsistent with economic survival, regardless of any personal desire to do so. Enterprises that lack economic viability eventually will lose control of their ecological resources to their competitors. In other words, farmers who can't survive financially will ultimately lose their farms to their profitable neighbors. Agriculture cannot be sustained if the only profitable farmers are those who degrade their agro-ecosystems, because such farmers will not be economically viable over time.

A fundamental purpose of public policy is to resolve conflicts between the short-run interests of individuals and the long-run interests of society as a whole. Ecologically sound systems of farming can be made economically viable through the public policy process. However, society ultimately must pay the costs of such policies, either through the price of food and fiber or through government taxing and spending. Such public policy may require nothing more than agreeing that no one has the right to pollute or degrade the natural environment or to exploit another person for economic gain. By one means or another, farming systems must be made economically viable as well as ecologically sound if they are to be sustainable. Neither is more important than the other.

The ultimate consensus that a sustainable agriculture must be socially responsible is still emerging. However, arguments that an economically viable and ecologically sound system of agriculture can be sustained in the absence of social equity and justice ignore fundamental human nature. At their very core, such arguments beg the question of sustainability *for whom*, or at least for how many and at what level. No set of ecologic possibilities can sustain the maximum population that humankind might possibly choose to reach on this earth. Nor is it ecologically possible to sustain even the current human population at any level of per capita resource consumption humanity might choose.

The history of human civilization provides nothing to support a hypothesis that either population or consumption—regionally or globally—will automatically adjust to optimum sustainable levels. To the contrary, overpopulation and unrestrained greed inevitably result in destruction and degradation of the natural resource base. Historical evidence from earlier civilizations suggests that this degradation will continue to a point where only a fraction of the population can be sustained that might have been sustained if overpopulation had been avoided. No set of ecological constraints will prevent starving people from consuming the seeds that might have produced a bountiful harvest, if the harvest was to come only after the people were dead.

Human societies that lack economic equity and social justice are inherently unstable, and thus are not sustainable over time. Such systems will be characterized by recurring social conflicts that can do irreparable damage to both the economic and ecological systems that must support them. Nothing in history indicates that human societies are any more resistant, resilient, or regenerative than are ecological communities. In an age of nuclear weapons and other forms of mass destruction, one instance of societal failure can destroy the ecosystem of an entire region. So-called natural phenomena such as deserts, droughts, floods, and famines are more frequently the result of failed social systems than of any naturally occurring catastrophe alone. Agriculture is a creation of human society that can be destroyed by human society.

An important dimension of human society is the ability to learn, discover new options, and to choose responses that are different from those of past civilizations. As far as we know, this ability to anticipate consequences that we have never experienced is unique to the human species. Sustainability simply is not possible unless we develop the collective will to exercise this uniquely human trait. An agriculture that meets the basic human food and fiber needs of society and promotes social equity and justice for all is no less essential for sustainability than is an ecologically sound and economically viable agriculture.

Some may question the wisdom of placing the burdens of global sustainability on American agriculture. One might logically conclude that American agriculture is only one part of global agriculture, and that agriculture is only one small aspect of the larger global ecosystem. If risks arising from lack of sustainability within American agriculture can be counteracted elsewhere within global agriculture, or within the rest of the global ecosystem, the system as a whole will still be sustainable. This conclusion is valid but only within limits.

As an analogy, the human body is a living system. The basic function of body organs such as the liver and kidneys is to handle wastes generated by other bodily functions. Some organs such as the heart and lungs may adjust their activity to accommodate stresses placed on them by other parts of the body. Waste generation is a normal function of any living organism, and some level of stress is necessary for a healthy body. However, the body as a whole is limited in its ability to assimilate wastes and absorb stress. When its critical limits are exceeded, the overstressed organ, a subsystem of the body, begins to die.

When a critical organ or part of the body dies, the whole body dies. The system ceases to function.

Likewise, when agriculture production in a particular field is not autonomously sustainable, it places stress on the farming system as a whole. When a farm is not autonomously sustainable, it places stress on the community of which it is a part. When the agricultural sector is not sustainable, it places stress on a nation, and a nation that is not sustainable places stress on the rest of the world. Some lack of autonomous sustainability should be considered normal, even necessary, for a healthy, interdependent global society. However, the stresses that any one element places on the system as a whole should be monitored and managed, in the same sense that stresses on the human body need to be monitored and managed.

It is no less important to monitor and manage the social stress an agricultural system places on farm families and others in rural communities than it is to monitor the economic stress agriculture places on food consumers or the ecological stress agriculture places on its natural environment. An agricultural system that destroys a critical element of an agro-ecosystem will degrade and eventually destroy the system as a whole.

We should be willing to ask of any proposed agricultural technology, enterprise, or activity, is it socially responsible? Competent, well-informed scientists will disagree on the answer. Such is the nature of science. Questions of social responsibility ultimately must be answered by society—by families, communities, and others who are collectively affected by agricultural decisions. However, it is logically imperative that we recognize that ecological soundness, economic viability, and social responsibility all are essential, and thus equally critical, to sustainability.

How do we turn these ideas into actions? I suggest that we do so by simply asking of every decision we confront, will the consequences be ecologically sound, economically viable, and socially responsible? We can then gather information that will give us the knowledge needed to answer this three-part question. We can never know for sure whether our conclusions or decisions are right or wrong, as sustainability is about *forever*. However, we will at least be asking the right questions, and by focusing our efforts on gathering the right information and pursuing the right knowledge, we should at least improve the odds of finding the right answers.

The foregoing thesis may not adequately define sustainable agriculture, but it does define an approach to working toward agricultural sustainability. The usefulness in defining such an approach may be made more apparent by speculating on who is likely to reject the approach and who is likely to embrace it.

First, those who do not accept agriculture as a legitimate human activity are likely to reject this approach, while those who question the sustainability of agriculture as it is currently practiced will likely embrace it. Those who believe that total elimination of commercial pesticides and fertilizers is absolutely necessary to achieve agricultural sustainability may reject this approach, while others who are more concerned with the overuse and misuse of agricultural chemicals may find it acceptable.

This approach will likely be rejected by those who contend that if a system is ecologically sound, then social values and economic incentives will adjust to ensure sustainability. It will also be rejected by those who contend that if a system is profitable, it's sustainable, period, as well as by those who contend that it is not necessary to use public policy to achieve economic sustainability. But it will be embraced by those who see the necessity for balance and harmony among the economic, social, and ecological dimensions of sustainability.

This approach may not be acceptable to those who see other living species as having as much right to the earth's resources as humans, but it may be supported by those who see human survival and well-being as critically interrelated with the other biological and physical elements of the global ecosystem. Thus, it likely will be rejected by those who feel that animals have rights but embraced by those who are dedicated to treating animals with respect and to preserving biological diversity.

This approach quite likely will be rejected by those who see agriculture as separable from the rest of the nature, who would separate the places where we farm from places we live and from places where we commune with nature. But it will be embraced by those who understand that we ultimately farm, live, and commune with nature in the same places because nature is inseparable.

In general, the proposed approach will be acceptable to those who would pursue a wide range of alternative means to achieve agricultural sustainability but will be rejected by those who see the alternative means for achieving sustainability as narrow or exclusive. While not providing a precise definition of sustainable agriculture, I hope that my thesis can at least provide a foundation from which such a consensus will evolve. And such a consensus will allow us to continue moving toward the common goal of agricultural sustainability.

! ! Presented at the Michigan Agriculture Mega-Conference, sponsored by the Michigan Agricultural Stewardship Association, Lansing, Michigan, January 12–13, 1996.

9

Foundational Principles of Soils, Stewardship, and Sustainability

The dictionary defines foundation as "the basis upon which something stands or is supported."[1] The basic premises of this discourse on "foundational principles" is that soil is the foundation for all of life, including human life, that stewardship of the soil is the foundation for agricultural sustainability, and that sustainability provides the conceptual foundation for wise soil management.

Virtually all living things require food of one kind or another to keep them alive. Life also requires air and water, but very few organisms can live on air and water alone. Many organisms that are not directly rooted in the soil—things that live in the sea, on rocks, or on trees, for example—still require minerals from the earth. They must have soil in some form from somewhere. Living things other than plants get their food from plants or from other living things that feed on plants, and plants feed from the soil. Some forms of life may seem to have no roots in the soil, but soil is still at the root of virtually all life.

I am not a soil scientist. I took a class in soils as an undergraduate and I have learned a good bit about soils from reading and listening to other people over the years. But I make no claim to being an expert on the subject. So I will try to stick to the things that almost anyone might know, or at least be able to understand, about soil.

As I was doing some reading about soil, I ran across a delightful little book called *The Great World's Farm*, written by an English author, Selina Gaye, at the turn of the twentieth century.[2] Back then, people didn't know so much about everything, so they could get more of what they knew about a lot more things into a smaller book. The book begins by explaining how soil is formed from rock, proceeds through growth and reproduction of plants and animals, and concludes with cycles of life and the balance of nature. But the book stresses that all life is rooted in the soil.

Initially, molten lava covered the earth's crust. So all soil started out as rock. Most plants have to wait until rock is pulverized into small particles before they can feed on the minerals contained in the rock. Chemical reaction with oxygen and carbon dioxide, wear by wind and water, expansion and contraction from heating and cooling, and rock slides and glaciers have all played important roles in transforming much of the earth's crust from rock into soil. However, living things also help create soil for other living things.

Lichens are a unique sort of plant that can grow directly on rock. Their spores settle on rock and begin to grow. They extract their food by secreting acids that dissolve the minerals contained in the rock. As lichens grow and die, minerals are left in their remains to provide food for other types of plants. Some plants that feed on dead lichens put down roots that penetrate crevices in rocks previously caused by mechanical weathering. Growth of roots can split and crumble rock further, exposing more surfaces to weathering and accelerating the process of soil making.

Specific types of rock contain a limited variety of minerals that provide food for limited varieties of plants—even when pulverized into dust. Many plants require more complex combinations of minerals than are available from any single type of rock. So the soils made from various types of rocks had to be mixed with other types before they would support the variety and complexity of plant life that we have come to associate with nature. Sand and dust were carried from one place to another by wind and water, mixing with sand and dust from other rocks along the way. Streams and glaciers also have been important factors in mixing soil. Some of the richest soils in the world are fertile bottomlands along flooding streams and rivers, loess hills that were blown and dropped by the wind, and soil deposits left behind by retreating glaciers.

Quoting from *The Great World's Farm*, "No soil is really fertile, whatever the mineral matter composing it, unless it also contains some amount of organic matter—matter derived from organized, living things, whether animal or vegetable. Organic matter alone is not enough to make a fertile soil; but with less than one-half percent of organic matter, no soil can be cultivated to much purpose."[3] After the mixed soil minerals were bound in place by plants, and successions of plants and animals added organic matter and tilth, the mixtures became what we generally refer to as soil.

The first stages of soil formation are distinguished from the latter stages by at least one important characteristic. The dissolving, grinding, and mixing required millions of years, whereas soil binding and the addition of organic matter can be accomplished in a matter of decades. Thus the mineral fraction of soil is a nonrenewable resource—it cannot be recreated or renewed within any realistic future time frame. But the organic fraction is at least somewhat renewable or regenerative, in that it can be recreated or renewed over decades or at least over a few generations. Misuse can displace, degrade, or destroy the productivity of both fractions of soils within a matter of years. And once the mineral fraction of the soil is lost, its productivity is lost forever.

If there are to be productive soils in the future, we must conserve and make wise use of the soils we have today. The soil that washes down our rivers to the sea is no more renewable than are the fossil fuels we are mining from ancient deposits within the earth. In spite of our best efforts, some quantity of soil will be lost—at least lost to our use. Thus, our only hope for sustaining soil productivity is to conserve as much soil as we can and to build up soil organic matter in order to enhance the natural productivity of the soil that remains.

In times not too long past, the connection between soil and human life was clear and ever present. Little more than a century ago, most people were farmers and those who were not lived close enough to a farm to know that their food came from the soil. They knew that when the soil was rich, when the rains came, and when the temperature was hospitable to plants and animals, food was bountiful and there was plenty to eat. They knew that when droughts came, plants dried out and died, the soil was bare, and there was little to eat. They knew that when the floods came, plants got covered with water and died, the soil was bare, and there was little to eat. They knew very well that their physical well-being, if not their very lives, depended on the things that lived in and from the soil.

William Albrecht, a well-known soil scientist at the University of Missouri during the middle of the twentieth century, hypothesized that people from different parts of the country had distinctive physical characteristics linked to the soils of the area where they grew up. He attributed those physical distinctions to differences in nutrient values of the foods they ate, which in turn depended on the makeup of the soils on which their foodstuffs were grown.[4] Albrecht's hypothesis was never fully tested. As people began to move from one place to another throughout their lives, and as more and more foodstuffs were shipped from one region of production to another for consumption, people no longer ate foods predominantly from any one region or soil type. But it's quite possible that when people lived most of their lives in one place and ate mostly food produced locally, their physical makeup was significantly linked to the makeup of local soils. Today, we eat from many soils, from all around the world. Still, as the saying goes, "You are what you eat." If so, what we become comes from the soil from which we eat.

Urban dwellers in particular have lost all sense of personal connection to the farm or the soil. During most of the twentieth century, many people living in cities either had lived on a farm at one time or knew someone, usually a close relative, who still lived on a farm, which gave them some tangible connection with the soil. At least they knew that *land* meant something more than just a place to play or space to be filled with some kind of commercial development. But most personal connections have been lost with the aging of urbanization. One of the most common laments among farmers today is that people no longer know where their food comes from. It is sad but in many respects true. The connection between soil and life is no longer so direct nor well understood, but it is still there.

What's even sadder is that many farmers don't realize the dependence of their own farming operation on the health and natural productivity of their soil. They have been told by the "experts" that soil is little more than a medium for propping up the plants so they can be fed with commercial fertilizers and protected by commercial pesticides until they produce a bountiful harvest. In the short run, this illusion of production without natural soil fertility appears real. As long as the soil has a residue of minerals and organic matter from times past, with only annual amendments of a few basic nutrients—nitrogen, phosphorus, and potash being the most common—crop yields can be maintained. Over time, however, as organic matter becomes depleted, production problems appear and it becomes increasingly expensive to maintain productivity. As additional "trace elements" are depleted, soil management problems become more complex. Eventually, it will become apparent that it would have been far easier and less costly over the long run if we had maintained the natural fertility of the soil. But by then much of the natural productivity will be gone forever. In the meantime, many farmers will continue to have little sense of their ultimate dependence on the soil.

Still, virtually all of life depends on soil. Life requires food, and for nearly all living things, no source of food exists other than the living things that depend directly or indirectly on the soil. This is a foundational principle of natural science, social science, and human health that should be taught at every level in every school in the world—beginning in kindergarten and continuing through college. That we must have soil to live is as fundamental as the fact that we must have air to breath, water to drink, and food to eat. It's just less obvious.

Soil is being eroded at rates far in excess of the rate of soil regeneration—in the United States and around the world. Experts may debate whether society can maintain agricultural productivity while losing soil, but there is no argument that humanity is losing useful soil at rates greatly exceeding the natural rates of new soil formation.

For example, U.S. government farm payments during the 1990s have been conditioned on "conservation compliance." An early proposal for conservation compliance was that soil loss must be limited to a rate of T, the soil loss "tolerance rate." T was defined as the estimated number of tons of soil that could be lost each year without reducing long-run productivity. Estimates of T included liberal assumptions about improvements in production technology based on past increases in crop yields, which resulted primarily from increased use of commercial fertilizers. T for most soil types, at least in the Midwest, was estimated at around five tons per acre per year.

I have heard estimates that soil can be regenerated at rates ranging from something less than one ton per year down to a fraction of a percent of a ton per year, depending on whether regeneration referred to renewing topsoil, the organic fraction, or to the total soil profile. Regardless, soil losses even at the rate of T would mean losses far in excess of reasonable rates of soil regeneration. The actual farm bill definition of "conservation compliance" allows erosion rates well in excess of T. Soils in the United States not covered by government programs, and land in many other countries of the world are eroding at still far higher rates, oftentimes essentially unchecked by any means of soil conservation.

Since soil is essential to all life, and since we are losing soil at rates greatly exceeding rates of soil regeneration, how are we going to sustain life on earth? The current answer seems to be that we will rely on future advances in production technology. In other words, we apparently have faith that future technological advances will stay ahead of net soil loss so that we can feed even more people better, or at least as many people as well, with less soil. But even under this assumption, we would eventually run completely out of soil. At some future time, we would be back to bare rock. Can we reasonably rely on future technology to feed the world from bare rock?

In the early days of the sustainable agriculture program, I was invited to make a statement regarding sustainability at a meeting in Washington DC. I stated that it should be obvious to anyone that soil conservation was an essential part of a sustainable agriculture because we simply could not sustain agricultural production without soil. After the meeting, a fellow came up and challenged my statement. He pointed out that production had continued to increase over the past hundred years or so, even though soil losses obviously had exceeded regeneration during this period of time, and there was no reason that this could not continue indefinitely. So I asked, "What's going to happen when there is no more soil to which to apply those technologies?" He answered, "By then, we will be able to grow our food in the sea." I replied, "Okay, let's say you are right, but what makes you think if humanity uses up all the soil it won't also use up all of the productivity of the sea? What will they do then?" He answered, "By then they will be able to grow food on other planets."

The point of repeating this dialogue is that if people have a blind faith in human technology—in the ability of people to conquer and remove all constraints of nature—then they are not concerned about soil conservation or soil regeneration. In fact, they are not concerned about the sustainability of agriculture, reliance on nonrenewable energy, pollution of the environment, social or economic inequities, or anything else that has to do with the long-run well-being of humanity. They have a blind faith that any problem created by humans can be solved by humans. We are solving problems today that were caused by past generations, so future generations should expect to solve any problems we create today. In their minds, those of us who don't share this faith in the future are just a bunch of pessimistic "Chicken Littles."

If this lack of concern for long-run sustainability were limited to the few individuals who openly express it, there would be no problem. But the passive belief in supremacy of "man over nature" seems to permeate much of society. It's reflected in the way most people work, play, and live, regardless of whether they have actually thought much about the future implications of their choices. They just assume that someone will always make more to replace whatever we use up or fix what we mess up, or if not, we won't actually need anything we destroy. Maybe they are right, but what if they are wrong? What are the odds that people of the future won't really need soil to live?

We need to understand that when we challenge people to stop and think about the long run, we are challenging their basic beliefs. Beliefs cannot be challenged with facts or logic. Our beliefs provide the foundation for our mental models and our worldviews. Facts have different meanings to different people depending on their specific mental models or the logic of their particular ways of thinking. Those mental models or ways of thinking are determined by beliefs concerning how the world works and where they think they fit within it. Beliefs cannot be proven, because all proofs depend on a specific set of beliefs. Thus, the only way to change the behavior of those who have a blind faith in technology is by challenging their worldviews, their beliefs, by asking them to rethink their assumptions concerning how the world works and their role within it.

Soils of the earth will be saved only through a growing sense of stewardship—among farmers, within communities, and around the globe. Conservation initiatives, including conservation compliance and public research and education, are all destined to fail in the absence of a strong sense of stewardship. We simply won't have the necessary government programs, or we won't enforce them, without a social commitment to stewardship. Even the best of conservation practices will not be widely used unless or until there is a consensus in support of human stewardship of the earth's resources.

Stewardship in general is defined as "the individual's responsibility to manage his life and property with proper regard to the rights of others."[5] Using this general definition, those with blind faith in technology might argue that regard for the rights of future generations is unnecessary, and thus, *no* regard is *proper* regard. So stewardship of soil, in the sense of conservation and regeneration, must be based on something deeper and more fundamental than an individual's personal definition of proper regard.

Most references to stewardship, at least in the United States, are linked to the Christian faith. In Genesis 1:26, God said, "Let man . . . have dominion over the fish of the sea, and over the birds of the air, and over the cattle, and over the earth, and over every creeping thing that creeps upon the earth." Genesis 9 begins, "and God blessed Noah and his sons, and said to them, be fruitful and multiply, and fill the earth." Unfortunately, some have used these scriptures to justify their right to exploit the earth and everything upon it.

But Jesus said in reference to stewardship, "Every one to whom much is given, of him much will be required" (Luke 12:48), and "As each has received a gift, employ it for one another, as good stewards of God's varied grace" (1 Peter 4:10). And in 1 Corinthians 4:2, Paul wrote, "It is required of stewards that they be found trustworthy." It seems clear that whatever form of dominion we humans have been given over the earth and other living things carries with it a responsibility to use those things for the long-run benefit of humanity, not just for ourselves and not just for the current generation.

Benefits of true stewardship do not accrue to the steward, either in total or in part.[6] They accrue to someone else, possibly some unknown person in some future generation in which the steward may not even have a direct descendant. Investments in true stewardship are simply not made unless one feels some spiritual responsibility for the future of humanity. Spirituality may be associated with one's religion, but it need not be. Religious practice is simply a means of fulfilling the responsibilities that one believes are inherent aspects of being a worthy human being. Religious or not, those without spirituality will see no logic in taking care of the soil for the benefit of future generations.

Paraphrasing noted religious philosopher William James, spirituality is a felt need to be in harmony with an unseen order of things.[7] First, for persons to be spiritual, they must believe that some higher, unseen order of things actually exists. Next, they must feel some responsibility or need to conform to that order and to allow it to influence and shape their actions. A spiritual person feels a need to live in harmony with fundamental laws of nature, including human nature, laws that define the higher order.

The noted conservationist Aldo Leopold expresses his spirituality quite simply in *A Sand County Almanac*: "Examine each question in terms of what is ethically and esthetically right, as well as what is economically expedient. A thing is right when it tends to preserve the integrity, stability, and beauty of the biotic community. It is wrong when it tends otherwise."[8]

A common thread of all expressions of spirituality is the existence of an unseen order that defines the oneness of all things. Rightness is defined as harmony with this oneness. It is wrong to create disunity or disharmony. We as people are a part of this higher order, of this whole. We may attempt to understand it and to have an influence within it, but we did not create it nor can we change it. Thus, we must seek peace through harmony, not dominance. In finding harmony with others around us, and for those of future generations, we find peace within ourselves.

Soil is formed and destroyed according to fundamental laws of nature that we have no power to change. When we conserve it, protect it, and maintain its natural fertility, we are working in harmony with nature—in harmony with the higher order of things. When we erode or degrade it, we are creating disharmony with nature—disharmony with the higher order of things. If we feel no spiritual sense of responsibility to take care of the soil, neither for future generations nor for any purpose other than our own narrow self-interest, we have no true sense of stewardship for the soil. A true sense of soil stewardship arises only from a sense of responsibility for others, to be trustworthy caretakers of the whole of life, to live in harmony with a higher order of things.

We humans have been given dominion over the soil of the earth. We can flush it down the rivers, poison it with salt and other chemicals, cover it with concrete, or farm it until it is worn out. Or we can conserve and protect it from erosion, build up its organic matter, treat its deficiencies, preserve it as farmland, and farm it in ways that will sustain its productivity indefinitely. Dominion empowers us with choices, but stewardship enriches us with responsibility. We can be good stewards of the earth only if we wisely choose to make investments for the benefit of humanity. The reward of true stewardship comes from knowing that we have done the right thing by acting in harmony with some higher order. The reward of true stewardship is harmony and peace within ourselves.

Soil stewardship is the foundation for a sustainable agriculture. A sustainable agriculture must be able to meet the needs of those of the present while leaving equal or better opportunities for those of the future. Sustainability applies the Golden Rule across generations: to do for those of the future as we would have them do for us.

An agriculture that meets the needs of both the present and future must be ecologically sound, economically viable, and socially responsible. All three are necessary. Stewardship relates most directly to the ecological dimension of sustainability. Ecologically responsible actions may be motivated by our own benefit, in protecting ourselves from environmental harm. Farmers presumably would not use chemicals in ways that destroy their health, poison their own food, or pollute their water supply. But the ecological dimension of sustainability relates even more directly to making ecological investments for the benefit of others. Sustainability requires that we consider the health and well-being of those downwind and downstream. Sustainability requires that we conserve nonrenewable resources—soil, energy, clean air, and clean water—for future generations. Thus, ecological sustainability is deeply rooted in a strong sense of stewardship—our responsibility to take care of things for the benefit of others.

The social dimension of sustainability relates to shared interests. In making social investments—giving some of what we have to benefit others—we may expect some personal rewards, but the rewards must be shared with others. We may make social investments for purely selfish reasons in that we expect our share of the benefits to be greater than our share of the costs. But we may also make social investments for purely altruistic reasons, in that we expect no direct benefit for ourselves but benefit only from fulfilling our responsibilities for others. Stewardship, in the sense of "to whom much is given, much will be required" and "as each has received a gift, employ it for one another," can be a powerful motivation for fulfilling our social responsibilities.

The economic dimension of sustainability relates to short-run self-interest. Economics does not deal with stewardship. Economic investments are made only when investors expect a positive return for themselves. An investment for the benefit of others—stewardship—is economically irrational. Others may benefit from our economic investments, but benefiting others is not an economically rational motive. Economic gain is never a logical motive for true stewardship.

The economic dimension is no less important than the social and ecological dimensions are in ensuring sustainability. Leopold said to consider the ethics and aesthetics as well as economics. One cannot be expected to take care of others unless they are able to take care of themselves. Treating others as we would like to be treated doesn't mean much unless we expect to be treated well ourselves. Economic viability is necessary if farmers are to maintain dominion over the resources for which they are to be good stewards. Or to put it bluntly, if a farmer goes broke, there is no way he or she can take good care of the soil.

Conflicts arise between economics and sustainability only because too often economics is allowed to dominate everything else. Economic viability is not the same thing as profit maximization. Sustainability requires a measure of profitability, but short-run profit maximization invariably leads to ecological degradation and social exploitation. Sustainability requires balance and harmony among all three—economic, ecological, and social.

Soil conservation quite likely is the most widely embraced of all sustainable soil management practices and has been a major public policy issue for most of the twentieth century—at least after the "dust bowl" days of the 1930s. If the soil is washed or blown away—to somewhere where it can't be farmed—it is no longer useful to agriculture. Since it takes hundreds of millions of years to replace the mineral fraction of soils, erosion obviously degrades the long-run sustainability of agriculture. However, interest in soil conservation obviously is not limited to those who identify with sustainable agriculture.

Conservation tillage, contour farming, strip cropping, terracing, and cover crops are all examples of farming practices designed to reduce erosion. These practices address the ecological integrity of agriculture—they keep the soil, the ecological foundation for farming, in place so it can be farmed. The social impacts of soil conservation are different for different practices. Conservation tillage that relies on bigger, more costly equipment, for example, continues the push toward fewer, larger farms. Practices such as cover crops and strip cropping may favor smaller, more management-intensive farms. Soil conservation practices also have economic benefits—particularly over the long run. But a lot of farmers apparently feel that the economic payoff is either too small or too long in coming, since it seems they must be bribed or coerced through government programs to protect their land from erosion.

The North Central Sustainable Agriculture Research and Education (SARE) program lists four different categories of research and education related to soil management: nutrient mineralization, soil microbiology, soil organic matter, and soil quality. The SARE program lists soil conservation as a natural resource issue.

Nutrient mineralization includes a whole range of issues related to the processes by which the mineral fraction of the soil is made available to plants. Interest in nutrient mineralization is not unique to sustainable agriculture. However, sustainable agriculture research and education tends to focus on mineralization as the primary process by which nutrients are supplied to crops, rather than reliance on routine replacement of nutrients through use of commercial fertilizers. Conventional industrial agricultural production systems may be viewed as linear, input/output processes in which fertilizers, pesticides, and so on are inputs and crops are outputs. Sustainable agriculture is oriented toward a cyclical process focusing on interrelated nutrient, water, plant, and energy cycles. Commercial fertilizers are viewed as amendments or supplements that can strengthen natural mineralization processes. Crops can collect and make nutrients available to companion and succeeding crops, sometimes through animal intermediaries, in addition to producing marketable commodities. Sustainable systems keep reliance on external inputs to a minimum.

Soil organic matter is a popular management topic among sustainable agriculture advocates—particular those who prefer organic farming methods. Organic matter is in the fraction of the soil profile typically referred to as topsoil. Although plants can grow without a whole lot of it, no soil is really fertile unless it contains some amount of it. And with few exceptions, soils with higher levels of organic matter are healthier, more productive soils.

Organic matter contains nitrogen, carbon, and other minerals, with proportions depending on the source. Nitrogen, a critical nutrient for plant growth, is the most important nutrient in organic matter that is not available to plants from the mineral fraction of soil. Nitrogen is abundant in the atmosphere, but most plants cannot take it directly from the air, legumes being the notable exception. Nitrogen can be added from commercial sources, but nature's way of supplying nitrogen is through soil organic matter.

Organic matter can be maintained by returning as much of the biomass as possible, from plant and animal sources, back to the land on which it was derived. Organic matter can be enhanced in one place by adding biomass that was derived from crops grown elsewhere or by growing crops that add more to the soil than they remove. Crop residues, compost, livestock manure, and green manure crops are all sources of organic matter that can be managed to enhance the natural productivity of soil.

Organic matter, while a popular topic in sustainable agriculture circles, seems to receive far less attention in the conventional agriculture community, among both farmers and scientists. Perhaps this is because commercial fertilizers are simpler and easier to use or because they are more profitable, at least in the short run. But there is little doubt that the organic matter that has accumulated over decades is being mined from many soils across the United States and around the globe. Organic matter is taken seriously among farmers associated with the organic and sustainable agriculture movements. Some farmers likely farm organically only because they are able to get an organic-market price premium that makes organic methods more profitable. But those who farm organically for philosophical reasons, and those who manage organic matter for sustainability, balance stewardship with economic considerations.

Soil microbiology relates to the living fraction of soil organic matter. The living fraction of soil contains a whole host of organisms ranging in size from bacteria to earthworms. Many of these organisms are involved in the process of turning crop residue and animal manure into soil humus—the nonliving fraction of soil organic matter. One of the most visible indicators of a healthy soil high in organic matter, other than color, is an abundance of earthworms. However, soil microbiology deals with the smallest of soil organisms, including those that contribute to healthy, productive soils as well as those typically identified as diseases or pests. The purpose of soil microbiology is to understand the role of these organisms, individually and collectively, in affecting the health and fertility of soil. The focus of soil microbiology in conventional circles seems to be on controlling specific diseases and pests, while the sustainability folks tend to focus more on maintaining healthy, productive soils.

Soil quality relates to the whole range of attributes that determine the inherent productivity of a soil. Managing for soil quality is the essence of managing soils for sustainability. Soils that are richer in minerals and higher in organic matter, with a healthier community of living organisms, tend to be more naturally productive than are soils lacking in these characteristics. Such soils not only provide more nutrients, they also hold more water and air and provide an environment in which roots can more easily reach out for nutrients and down for water. Consequently, plants that grow on high-quality organic soils tend to be less subject to stress, and thus tend to have fewer insect and disease problems. A common recommendation in sustainable agriculture circles is to "manage the soil instead of the crop."

Threats to the long-run sustainability of the soil resource almost invariably arise from short-run economic considerations. Nutrient management with commercial fertilizers and soil amendments is simpler, easier to standardize, and typically less costly in the short run. So it's easier for a farmer to farm more land more profitably in the short run by using commercial fertilizers rather than maintain natural fertility through organic or sustainable approaches to soil management. As more farmers have adopted the input/output industrial approach to soil nutrient management, total agricultural output has expanded, prices have dropped, and profit margins have been squeezed. Those who still manage for soil quality find themselves under increasing economic pressure to mine their soils for short-run profits rather than maintain soil quality for long-run economic viability. They have to survive the short run in order to be sustainable in the long run.

Additional problems arise from the separation of landownership from actual farming. Restoration and maintenance of soil quality require continuing investments in future productivity—and these investments may be substantial. If farmers don't own the land they farm, how can they be sure their economic return on their investments in soil quality will not accrue to someone else, specifically to the landlord? If instead the landlords want to invest in soil quality, how can they be sure that their renters, the farmers, will not mine the soil for maximum year-to-year profit rather than help build the soil for long-run sustainable yields?

The aging farm population presents still another challenge to sustainable soil management. Farmers who own land when they are young might reasonably expect to reap the economic rewards from investing in soil quality over much of their tenure on the farm. But as they grow older, they have fewer years over which to recoup the returns from their earlier investments. As the average age of farmers, already well over fifty, moves higher, it becomes increasingly difficult economically to justify investments in soil quality.

The corporatization of agriculture represents a threat to the soil far greater than we have yet seen. A publicly held corporation can allow no motive to take precedence over profit and growth. Agriculture of the future quite likely will be dominated by such corporations. For them, the long run may be two to four years, and the only thing that really matters much is their quarterly dividends and the week-to-week price of their stocks. Their stockholders demand nothing else. Farmers who become corporate contract producers will be forced to conform to corporate standards of "soil stewardship," which means no more than whatever is absolutely required by law. And farmers who are convinced that they will be forced out of business by the corporations will have an economic incentive to mine their soil before they sell out and leave.

So where is the hope for humanity? All of life ultimately is rooted in the soil. If we destroy the productivity of our soils, we destroy the foundation for life, and ultimately we destroy ourselves. The hope for humanity ultimately resides in each of us. We are a part of the web of life that includes all things on earth. Whatever we do or don't do affects the web of life, including every living thing on earth.

Hope for the future of humanity is in each of us. To realize this hope we must proclaim to all who will listen, by all means at our disposal, that all of life depends on the soil. We must proclaim openly and without hesitation that we simply will not allow short-run economic considerations to take precedence over soil stewardship. The economy is a creation of humanity for the good of humanity; we cannot allow it instead to destroy humanity by destroying the soil. The market is not God. The economy is a human creation. We can, and we must, bend it and shape it to serve our needs, and not be enslaved by it. The long-run sustainability of human life on earth depends on a balance between economic, ecological, and social concerns, not on maximizing any one to the neglect of the others. It is our responsibility to restore this balance.

We must be willing to defend the principle of stewardship as a fundamental human responsibility, not simply a religious prerogative that one may or may not chose to exercise. Sustainability ultimately depends on a national and global consensus that stewardship of the soil must take precedence over short-run human greed. Given such a consensus, all would be compelled to practice stewardship, and none would suffer socially or economically as a consequence.

Research and education concerning sustainable soil management are necessary to support those who are willing and able to practice soil stewardship without a public consensus concerning its rightness. However, the sustainability of our soils, and thus the sustainability of life on earth, will not be assured until we reach a global consensus in support of soil stewardship. We can contribute foremost toward reaching this consensus by doing all we can as individuals to practice our commitment to true stewardship in all aspects of our lives. The principles of stewardship provide the sustainable foundation on which wise management of soils must be based.

:: Presented at the Southern Regional Train-the-Trainer Workshop, Sustainable Agriculture Research and Education Professional Development Program, Raleigh, North Carolina, September 22–24, 1999.

10

Economics of Sustainable Farming

For more than fifteen years, I taught the conventional principles of farm economics through various on-campus and extension courses at three different land-grant universities. I taught
farm management, marketing, finance, farm policy, and other
such subjects in an effort to help farmers maximize profits from
their limited economic resources. During the past fifteen years,
however, I have taught a different kind of farm management.
I have studied and taught the principles of sustainable farm
economics—the economics of sustainability. I am unable to
pinpoint the time of my transition from one to the other, as it
didn't happen all at once. For me, the transition began during
the farm financial crisis of the 1980s. During that time, I began
to realize that as a teacher of conventional farm economics, I
had done more to help create the farm crisis than I had done
to prevent it. The solution to the financial problems of farmers
would require something quite different from the farm economics I had been teaching.

First, I had to try to understand why the old economics of
farming wasn't working in the 1980s and wasn't going to work
in the future. I believe farmers, and those who work with farmers, must come to a similar understanding. We must understand why farmers have done the things they have done, and
why it hasn't worked, before we can understand why we need a
new economic approach and what the new economics of farming needs to be.

Conventional farm economics focuses on profit maximization. The underlying assumption is that more profit results in
a higher economic standard of living, which in turn translates

into a higher level of satisfaction and a higher quality of life. The discipline of economics may be characterized as a study of the optimum allocation of *scarce* resources among competing ends, so as to achieve the highest possible level of satisfaction or quality of life. Scarcity in economics means that economic resources are never sufficient to fully satisfy insatiable human wants and needs. Resources must be allocated or rationed among competing uses by putting resources to their most profitable use.

In economics, the final arbiter of value is the consumer. Resources are allocated in such a way as to best meet the wants and needs of people as consumers; producing is but a means of earning the privilege of consuming. Economics also assumes that markets are economically competitive, that any excess profits of producers will be quickly competed away and passed on to consumers. Thus, as farmers maximize their profits—so they can consume more—they are simultaneously allocating resources in such a way as to maximize the efficiency of resource use—so consumers can consume more. It all fits together very nicely under the assumptions of an economically competitive, capitalistic economy.

Maximum economic efficiency means maximum economic value relative to economic costs. In order for something to have an economic value, it must be scarce so that market prices can allocate its use among competing ends. Air and water, for example, have great *intrinsic* value, in that they are essential for life, but they have no economic value under most circumstances because they are not sufficiently scarce to command a market price. The typical water bill reflects the cost of delivering water but not an actual cost for the water. Only when clean air and clean water are made scarce through pollution or overuse do they take on an economic value.

Things that have no economic value are not counted in evaluating the economic efficiency of farming. Thus, conventional farm economics provides no incentive to protect the quality of air or water, or even the quality of soil, beyond maintaining its immediate productivity. Farm economics is about maximizing the economic value of the things produced and sold relative to the economic costs of the things used to produce them.

Although I didn't realize it in my earlier years, pursuit of economic efficiency greatly narrows the premise concerning the best approach to maximizing profits, as well as the best approach to maximizing quality of life. The pursuit of profits has pushed farmers toward an industrial model or approach to farming. The industrialization of agriculture has resulted in increased economic efficiency, but it has diminished the quality of life for many farmers, both those who have been forced out of business and the many who continue to farm. Industrialization also raises many questions concerning the impact of agriculture on the natural environment, on food safety and quality, and on societal quality of life.

Farmers abandoned diverse farming systems, which generally included both crops and livestock enterprises, in favor of more specialized farming systems. By specializing, farmers could become more efficient by doing fewer things better. A specialized cattle feeder, for example, could put on more pounds of beef per pound of feed than could a diversified farmer who fed out a few cattle in addition to doing a lot of other things. Farmers also discovered that gains from specialization could be enhanced by standardizing the functions involved in various production processes. Standardization allows each of the specialized functions to fit together more effectively in order to achieve maximum efficiency. For example, if more ranchers produced the kind of fall calf that would do well in winter stocker operations and would finish out more efficiently in the commercial feedlots, the whole process of beef production could be greatly improved.

As farming systems became more specialized and standardized, each function became more simplistic and routine. Many functions could then be mechanized—allowing predictable, reliable machines to replace often unpredictable and unreliable draft animals and human laborers. Commercial fertilizers, pesticides, hormones, and antibiotics made production processes more predictable, reliable, and repetitive. The mechanization and simplification of farming allowed each farmer to farm more land, use more capital, and supervise more workers. Consolidation of decision making has allowed farmers to achieve many of the economic efficiencies of large-scale industrial production.

As a consequence of industrialization, American agriculture has become one of the most productive and efficient agricultural economies in the world—at least in terms of the economic value of food and fiber relative to the economic costs of production.

This industrial approach to farming also led to some rather narrowly focused farm management strategies. Enterprise analysis has been one of the fundamental building blocks of conventional farm management. Farmers are told that they must be able to separate the farming operation into individual enterprises and calculate the costs associated with each economic enterprise, if they are to be able to manage for maximum profits. Also, they are told that they must separate costs and returns of the farm business from costs and returns associated with other family activities. The sole function of the farm is to provide income for the family, thus family and farm resources, costs, and returns must be kept separate. Analysis, by definition, means to take something apart, to separate it into its component pieces, in order to understand the whole through careful examination of the parts. Enterprise analysis is designed to help farmers understand the farm and to achieve higher farm profits, by taking the farm apart, piece by piece.

Farmers are asked to allocate their total costs among their various farming enterprises, at least to the maximum extent possible. The easiest costs to allocate are the variable costs, which include costs that will not be incurred unless a particular enterprise is carried out during a particular time. Variable costs also vary with the planned level of production of that enterprise. Feed, feeder animals, seed, fertilizer, and agrochemicals typically fall in the variable cost category. Fixed costs, on the other hand, are costs that will be incurred regardless of level of production or even whether anything at all is produced, and thus are more difficult to allocate among enterprises. For any given year, fixed costs include such things as buildings, equipment, and land. Farmers are even encouraged to calculate charges for labor provided by themselves and their families and to charge the farm for their management, although they may not pay themselves or their family actual wages or salaries. All such costs are to be allocated among the various farm enterprises.

The total costs associated with each enterprise are then compared with the projected market value of expected production to determine an expected net revenue or profit from each enterprise on the farm. The expected profit for the total farm is simply the sum of the profit estimates for the individual enterprises. Any costs that cannot be allocated to specific enterprises may be called "overhead costs" and deducted from the sum of enterprise profits.

The unspoken assumption of enterprise analysis is that the farmer will be able to identify specific enterprises that are contributing the greatest returns per unit of the farm's most limiting resource—be it land, capital, labor, or management. Farmers can increase profit only by increasing the efficiency with which they use the most limiting of the fixed resources. If the farmer has more than enough capital and labor to produce more of a crop but lacks additional land, for example, then land is the limiting resource. By increasing profits per acre, the farmer then can increase profits for the whole farm.

Thus, by shifting land, labor, capital, or management from enterprises in which the farm is least efficient to those in which the farm is most efficient, total farm profits can be increased. In addition, increased specialization may allow the farmer to achieve added efficiencies through utilization of larger or more specialized buildings and equipment, standardizing production processes, and thus allowing an increase in scale of operation. The result is increased profits through increased industrialization—that is, through specialization, standardization, and larger-scale operation.

However, increased specialization often results in increased risks. By specializing in one crop or a few crops, a farmer becomes more vulnerable to a crop failure due to weather or pest problems, or to depressed market prices for any of the crops produced. By specializing in one species of livestock, or one phase of production, a producer likewise is more vulnerable to disease or causes of poor performance or a cyclical downturn in prices. Thus as a farm abandons diversity and becomes more specialized, it becomes more vulnerable to both production and market risks.

In addition, specialization tends to increase financial risks. Financial risks are related to the ability of the farm to meet its debt repayment commitments. Financial risks are linked to production and market risks, since either low yields or low prices may cause the farm to suffer large losses and thus be unable to meet its financial commitments. Financial risks reflect the probability that the farm will lose more than it can afford to lose in any given year. Farms that rely more on purchased inputs, such as seed, feed, fertilizer, and chemicals, than on inputs produced on the farm increase the amount of out-of-pocket costs that must be paid up front, or at least at harvest time. As they increase investments in larger or more specialized buildings and equipment, farmers often borrow money that must be repaid on a regular basis. To cope with large financial risks, farmers with specialized, high-input, high-investment farms tend to rely on government programs to protect them from market and production risks. Early in my career, I spent a good bit of my time with conventional farmers talking about government price-support programs and government-subsidized crop insurance.

Industrial farmers are price takers in the marketplace. They produce standardized commodities and thus have no influence over the prices they receive. One farmer's U.S. No. 1 hard red winter wheat is the same as another's as far as buyers are concerned, which means that no farmer can get a higher price than any other. Prices vary over time with changing supply and demand, but commodity producers are price takers, not price makers. The only marketing decisions conventional farmers make is to decide when to establish a price for the things they produce. They may use forward contracts or options, or hedge using futures markets to manage price risks. By using such tools, farmers can price at some time before delivery or at delivery, and in some cases, they can defer pricing until after delivery. They also may set a specific price, a price range, or a minimum price. But the price is always one that is offered by the market; the farmer can only take it or leave it. In the face of such risks, some farmers resort to comprehensive production contracts that promise a fixed amount of return per unit of production. The farmer in essence becomes a landlord or a contract laborer for an agribusiness corporation.

Much of my early career as an extension economist was spent helping farmers maximize profits by managing the various types of risks associated with specialized large-scale production of standardized commodities. Management of this type of farming is no different in concept from management of any other industrial corporation.

During the farm financial crisis of the 1980s, I began to realize that the industrial approach to farm management was driving farmers out of business. Every time we helped some farmers improve their profits by specializing, mechanizing, and increasing their scale of operation, we were helping to force other farmers out of business. Our new technologies and management techniques were expanding farmers' ability to produce far faster than consumer demand was expanding for the things that farmers produced. Thus, farm profit margins grew narrower with each new round of technology, and each farmer had to increase the size of his or her operation just to survive—to spread management across more land, using more capital and more hired labor. As the farms grew larger, they obviously grew fewer in number. And with each new round of technology, some farmers had to fail so others could "succeed." There was no logical end to this process. This type of farming was not sustainable—at least not for farmers.

The industrialization of agriculture also made farmers more vulnerable to the chronically recurring periods of surplus production, initiated by good weather or good prices but made possible only by farmers' increasing ability to produce. Industrial agriculture meant high-capital agriculture, and much of the capital used by farmers during the 1980s was capital borrowed during the export boom years of the 1970s. A high-input agriculture meant a high-variable-cost agriculture, since direct cash cost of inputs made up a larger portion of total costs. Thus, when farm commodity prices plummeted during the 1980s, farmers were caught with large cash commitments, both for loan repayment and input costs, with little income from which to pay those costs. Their own resources—labor, management, owned land, and capital—had contributed very little to their farming operations. These farmers didn't have the option of simply taking less money out of the farming operation to pay themselves. The banker and the input suppliers demanded that their accounts be settled in full, regardless of the farm financial situation. Farmers who had done the things we "farm experts" had encouraged them to do in the 1970s were losing their farms in the 1980s. Something was fundamentally wrong with this kind of agriculture and this kind of farm management.

As I began to rethink the economics of agriculture, I became aware that the economics of industrialization not only encouraged farmers to exploit each other but also had encouraged them to exploit the land. Soil erosion rates had risen dramatically during the 1970s, as farmers farmed "fencerow to fencerow" and then tore out and farmed the fencerows. By the late 1980s, soil erosion had become a major national agricultural policy priority, as was reflected in the various conservation provisions of the 1990 Farm Bill. Commercial fertilizers and agrochemicals, necessary to support industrialization, also had raised serious questions concerning the quality of water in underground aquifers and streams. Organic farmers went to Washington DC in the mid-1980s with demands that USDA support their more ecologically sound approach to farming. And by the early 1990s, many more people among the general public were beginning to demand a more ecologically sustainable approach to farming.

The impact of agricultural industrialization on the social fabric of rural areas rose to the public consciousness as rural communities began to feel the brunt of the farm financial crisis of the 1980s. Once prosperous farming towns withered and decayed as large numbers of farm families were forced off the land. The land was still farmed, but there were fewer people to buy the things that supported local businesses. In addition, the larger industrial farms often bypassed the rural communities in order to save a few dollars on input costs or to get a few more dollars out of their products. Ultimately, the corporate takeover of hog farming, with their giant confinement feeding operations, or "hog factories," raised public awareness of the industrialization of agriculture and its destruction of the social fabric of rural America.

Today, farmers, rural residents, and society in general are demanding a more socially responsible, ecologically sound, and economically viable system of farming. This is the challenge of farm economics as we enter the twenty-first century: to help farmers build a more sustainable agriculture. Farms of the future must be economically sustainable, for farmers as well as consumers and society in general, but the profitability of farming cannot be sustained through exploitation of the land or exploitation of other people. To be economically sustainable, agriculture must conserve and protect the natural resources on which its long-run productivity depends. To be economically sustainable, it must contribute to the social and cultural quality of life for farm families and rural residents as it provides an adequate supply of safe and healthy food and fiber for society in general. Economic sustainability demands a new approach to farm economics.

Farming for economic sustainability begins with rethinking the basic purpose of farming. The only justification for maximizing profits in the old farm economics was that maximum profits were assumed to result in the highest attainable quality of life, both for farmers and for society as a whole. However, such an assumption is no longer defensible in light of the recurring farm financial crises of the last half of the twentieth century and growing evidence of the negative ecological and social impacts of agricultural industrialization. Economic well-being is a necessary dimension of quality of life, for farmers and for society, but it is not sufficient to ensure a desirable life of quality. We must pursue our economic quality of life by means that do not diminish our social and ethical quality of life in the process.

Profit maximization is a reflection of the natural pursuit of our individual self-interest. This is perhaps the single most appealing premise of conventional economic thinking. However, concern for others is also a natural consequence of being human. We are social animals—we need positive relationships with other people, not only to be successful but also to survive. Thus positive relationships with other people contribute to our quality of life, regardless of whether we receive anything that contributes to our individual self-interest. Ethics and morality also are fundamental characteristics of being human. Most us believe in some higher power, or some higher order of things, from which we derive purpose and meaning for our lives. Thus, moral and ethical behavior, including stewardship of the natural environment, contributes to our quality of life, regardless of whether such acts contribute to our individual, material self-interest.

Conventional economic thinking has led us to believe that we best serve the interest of society by pursuing our individual self-interest. However, the assumptions on which this proposition is based are no longer valid. The necessary conditions for competitive capitalism—sufficient buyers and sellers so that no single one can affect the market, perfect information concerning price and performance of products, freedom of entry into profitable enterprises and exit out of unprofitable ones, and the sovereignty of consumer tastes and preferences—no longer exist in today's economy. Large corporate entities dominate virtually every sector of the economy. It is neither easy to get into or out of most businesses today because of large capital requirements and all sorts of patents and copyrights. Billions of dollars spent on advertising designed to bend and shape consumer preferences make a mockery of assumptions of perfect information and consumer sovereignty. We no longer have a competitive, capitalistic economy—not in agriculture or anywhere else.

If we are to be socially responsible, we must make conscious, purposeful decisions to build positive relationships with other people. Thankfully, most people realize that the quality of their own lives is enhanced when they share with other people. The Golden Rule, "do unto others as you would have them do unto you," is a fundamental principle that underlies nearly every enduring religion and most of the major philosophies of the world. Humans have learned that their lives are made better by their acts of sharing with others. A socially responsible life is a quality life.

Conventional economic thinking treats the natural environment as something outside or external to the pursuit of self-interest. The environment represents a constraint to profit maximization. In reality, stewardship of the environment is an ethical or moral issue. Pursuit of individual self-interest may cause us to avoid doing anything to the environment that threatens our own health or the health of our loved ones. However, we will not conserve and protect resources for the benefit of future generations unless we believe stewardship to be a moral or ethical responsibility that gives purpose and meaning to our lives.

People of future generations can't participate in markets, so markets don't reflect the economic value they might place on resources. People of future generations can't vote in elections, so they can't shape public policy to protect their interests. But many moral and ethical people willingly choose to take care of the natural environment for the benefit of future generations because stewardship contributes to a desirable quality of life. A life of ecological integrity is a life of quality. We are just beginning to learn that the Golden Rule applies not only to others around us today but also to future generations.

The first principle of sustainable farm economics is the pursuit of *enlightened* self-interest, which recognizes the individual, interpersonal, and spiritual dimensions of self. This principle is reflected in nearly all of the most popular postindustrial approaches to farm management, including holistic resource management, biodynamic farming, permaculture, and organic farming.[1] The three-part goal of holistic management—forms of production, quality of life, and future landscapes—is just a different way of stating the economic, social, and ecological dimensions of sustainability. Biodynamic farming is about feeding the spirit as well as the body. Permaculture is about building a permanent, sustainable agriculture to support a permanent human society. The purpose of true organic farming is to support a permanent society, as much a philosophy of life as a means for making a living. In all these approaches to farm management, economic objectives are balanced with social and ecological objectives. The overall goal is to achieve a higher quality of life through harmony and balance among things economic, ecological, and social, rather than through maximization or minimization of anything.

The second principle of sustainable farm economics is to take a holistic approach to farm management. Rather than taking the farm apart piece by piece, the farm is considered as an indivisible, interdependent whole. In a sustainable farming operation, the relationships among the various components of the farm are as important as the components themselves. Traditional enterprise analysis tends to ignore, or at least to distort, the contribution of positive relationships to whole-farm economics. For example, when individual crops in rotations are evaluated separately, and when livestock enterprises are evaluated separately from crops, the potential for positive interrelationships among various crop and livestock enterprises is ignored. For example, crop and livestock enterprises can be integrated to manage pests, maintain soil health and fertility, efficiently utilize available labor, and diversify production and market risks.

Holistic management requires that the potential impact of changes in one or more enterprises be evaluated in terms of their impact on the economics of the whole-farm system. The various postindustrial approaches to farm management each advocate somewhat different methods of whole-farm management, but they all achieve the same basic end; they consider the farm as a whole rather than as a collection of enterprises. The fundamental question is how best to synthesize a whole farm or how best to put together an effectively integrated whole-farm system, rather than how to choose the best collection of individual enterprises. With holistic management, productivity is achieved through synergy, through building wholes that are greater than the sum of their parts.

One basic approach to whole-farm evaluation is closely related to "partial budgeting," a popular tool of conventional farm managers. In partial budgeting, a change in a specific enterprise or activity is evaluated by estimating its potential impact on the overall farm operation. It is called partial budgeting because only those aspects of the farm that will be affected by the change are budgeted. First, expected additions to income from the new enterprise are added to any expected reductions in costs in other enterprises as a consequence of adding the new enterprise. Next, expected additional expenses associated with the new enterprise are added to any expected reductions in income in other enterprises associated with new enterprise. Finally, the sum of the additions in costs and reductions in income are subtracted from the expected increases in income and reductions in costs to derive a net change in whole-farm income as a consequence of the proposed change.

If such a process is carried out carefully, the result should provide a reasonable estimate of the economic consequences of changing any part of a farming operation with respect to the farming system as a whole. The same process can also be followed to assess the social and ecological implications of changing any aspect of a farming operation. Budgeting would have to include such intangibles as amounts and quality of time available to spend with family and engage in community activities. Partial budgeting could also be used to assess the potential impact of changes in the overall farming operation on environmental stewardship, including such indicators as soil erosion, water quality, and biological diversity. Such an approach invariably must consider the family or the person farming as a part of the overall farming system. Factors such as availability of family labor and management, alternative uses of time, ethical and moral values, and the expression of family preferences and values through the farming operation ultimately must be balanced with economics in farming for quality of life.

Another principle of sustainable farm economics is strength through diversity. Biological and economic diversity are essential in building ecological systems that are durable as well as productive. Production, marketing, and financial risks can all be managed through diversity. In managing biological diversity, some important considerations include selecting a combination of crops and livestock enterprises—spatially, sequentially, and temporally—in order to break pest cycles or manage pest populations, maintain soil health and fertility, and efficiently utilize available resources. By relying on diversity rather than off-farm inputs to maintain productivity, farmers also reduce their out-of-pocket variable costs.

However, diverse systems typically require more labor and management, which often are committed and thus fixed in nature. But even if farmers increase their fixed costs, they may reduce their variable costs as they substitute labor and management for off-farm inputs. Thus, farmers can significantly reduce their financial risks by relying less on off-farm, purchased inputs and more on on-farm, owned resources, even if their total costs remain essentially unchanged. On such farms, most short-term losses due to adverse weather or markets can be absorbed by accepting a smaller return for labor and management. High-input, high-variable-cost farms are more vulnerable to the risks of economic failure than are low-input, high-fixed-cost farms.

In managing economic diversity, the most important considerations are to select combinations of enterprises that will tend to have offsetting patterns of market prices, so that profits from some enterprises will tend to offset losses from others. Even commodities with unrelated or uncorrelated price patterns add economic diversity. For example, a farm with four equal-sized enterprises with unrelated price patterns of equal variability will have only half as much income variability as a farm of the same size that specializes in only one of the four enterprises. However, diversity is not the same thing as variety. If different enterprises have the same basic production and market patterns, such as corn and soybeans, variety will do relatively little to reduce economic risks. Sustainable farm economics requires effectively integrated, economically diverse farming systems.

The final principle of sustainable farm economics is *individuality*—specifically, giving customers full economic value. Farm profitability simply cannot be sustained by selling undifferentiated farm commodities such as corn, hogs, cattle, or wheat in global markets dominated by large agribusiness corporations. Profits can be sustained only by providing customers with food and fiber products that are fundamentally different from the products they find in the supermarkets. This is perhaps the most difficult aspect of sustainable farm economics because it is the biggest stretch from traditional farm management. However, corporatization of agriculture has resulted in an agricultural sector in which individual farmers will not be able to compete, even if they are competitive in terms of price and quality. The corporations have sufficient power in the marketplace to deny market access to farmers who are not willing to sign comprehensive production contracts and settle for the role of landlord or contract laborer. Competing in commodity markets is no longer a matter of efficiency but rather of market power.

Food corporations must mass-produce and mass-market food products in order to achieve the economies of scale necessary to be competitive in today's global food system. As a consequence, most agricultural products in the supermarket today were selected far more for their adaptability to machine harvesting, efficient processing, transportability, and shelf life than for taste, tenderness, or nutrition. In addition, mass-produced foods must be targeted to the most common consumer tastes. The economic savings derived from mass production come from standardization, not from variety. But we don't all have the same tastes and preferences, and thus we value different food items differently. Sustainable farmers must give more consumers more of the things they value most rather than try to compete on cost or convenience.

Farmers who sell directly to customers in local markets have an opportunity to select crop varieties or livestock breeds for superiority in taste, tenderness, healthfulness, and nutrition rather than handling, transportation, and shelf life. They can sell products harvested at their peak quality and delivered fresh to local customers. Such advantages simply cannot be duplicated by industrial production systems, thus giving local farmers a sustainable market advantage.

Equally importantly, sustainable farmers can market their products based on their commitment to social responsibility and ecological integrity. Many consumers really do care where their food comes from, how farmers treat the land and the animals that provide the food products, and whether or not farmers are committed to making the world a better place. Study after study has shown that many people will pay a premium for food produced in ways they consider more sustainable. Industrial organizations may make claims of sustainability, but the industrial paradigm simply cannot meet the social and ecological standards of sustainability. Sustainable farming will require a different kind of marketing—one that gives their individual customers more value.

Perhaps the greatest challenge of economic sustainability for farmers is also its greatest potential reward. In order to sustain the profitability of farming, farmers must develop meaningful relationships with their customers. In order to sustain such relationships, farmers and their customers must know and trust each other. They must be committed to working together for their mutual good. Such relationships need not be limited to local residents, but farmers must view their customers as real people rather than as impersonal markets. A person can have a personal relationship with another person halfway around the world, but an agribusiness corporation can't have a personal relationship with anyone. A corporation is not a person. Personal relationships cannot be mass produced, so they can't be industrialized. But perhaps more importantly, the relationship between farmers and their customers can be one of the most important aspects of finding a more desirable quality of life through farming. And by sharing their commitment to stewardship of the natural environment, farmers and their customers can help each other to lead more purposeful and meaningful lives.

This kind of farm economics is different from the economics I taught to farmers in the 1970s and 1980s, but this kind of economics makes a lot more sense. It may require more work and certainly a lot more thinking, but it is a better way to farm and to live. There is no guarantee that this kind of farm economics will work for any given farmer or even for farmers in general. But the economics of sustainability provides a lot more hope for the future of farming than does the economics of industrialization. Hope is not the expectation that something will succeed or even that the odds of success are in your favor. Hope simply means that something better is possible. I know sustainable farming is possible and I know it would be better, even if the odds are against it. In the possibility of sustainability, there is hope. The farm economics of sustainability is a hopeful economics.

⁞ Presented at "Systems in Agriculture and Land Management," the annual conference of Holistic Resource Management of Texas, Fort Worth, Texas, March 2–3, 2001.

11

The Renaissance of Rural America

Most rural communities in America were initially established for the primary purpose of utilizing the natural resources located in rural areas. Natural resources—such as land, minerals, landscapes, and climates—must be utilized, at least initially, in the geographic locations where they exist. So people must move to where the resources are if society is to benefit from their use. Of course, the Native Americans were already using the resources of the land, but for purposes quite different from what the Europeans had in mind. The new settlers traveled west, dispersing themselves across the American countryside in patterns that seemed most appropriate for the natural resources they sought to exploit.

Some early American settlements were mining or logging communities, but the historic purpose of most communities in the United States was to realize the economic value inherent in agricultural lands. Distances between community centers tended to reflect the time it took farmers and ranchers to travel into town to trade their surplus production for necessary supplies. But the density of population in the European American settlements was determined largely by the number of farmers or ranchers needed to realize the perceived benefits from utilizing the land. The rangelands of the West were sparsely populated because one rancher could manage a herd of cattle roaming over hundreds, even thousands, of acres. Areas suited for truck farming and dairy operations were more densely populated because of the high human-input requirement for those enterprises. The Midwest was covered with diversified family farms with a corresponding rural population density.

Nonfarm economic activity in rural communities was closely related to numbers and types of farms. More service activities were needed in areas with larger farm populations. More people needed more health care, education, and other social services. Business activities in rural communities were closely related to the nature of farming enterprises and associated needs for markets and farm inputs such as credit, machinery, feed, and fuel. Rural service communities evolved into trade centers as early farmers moved away from self-sufficiency and began to specialize and trade among themselves. Many rural communities later became agribusiness centers as more people left nearby farms for urban areas and as the remaining farmers became more reliant on mechanization, markets, and purchased inputs. Reduced travel times also contributed to the growth of emerging trade-center communities at the expense of surrounding communities.

Over the past fifty years, many rural communities have lost their purpose. The trend during this period has been toward fewer, larger, and more specialized farms. The result has been declining rural populations, declining demand for local markets and locally purchased inputs, and a resulting economic decay of many rural communities. Some communities attempted to diversify their economy to reduce their dependence on agriculture, and others abandoned agriculture entirely as a source of economic development. Industry hunting became a preoccupation of many small town councils and chambers of commerce. Jobs, any kind at any cost, seemed to be the primary development objective in some declining rural communities. The lack of a geographical foundation to sustain new development was given little if any consideration.

Many development activities, lacking a geographic foundation, were rooted in nothing more than short-run exploitation of undervalued human and natural resources. Manufacturing jobs, often paying low wages, were expensive to attract and retain, although each company provided a relatively large number of jobs. The number of working poor—workers with full-time jobs who live below the poverty line—in rural areas has continued to rise. In addition, many manufacturing companies and branch plants that initially relocated in rural areas eventually moved to other countries where laborers are willing to work even harder for far less money. Efforts to attract low-quality, low-paying jobs are increasingly regarded as expensive and ineffective strategies for rural economic development.

Some new rural economic activities, such as tourism, vacation homes, retirement communities, and rural residences, may have strong geographic and economic foundations in climate, landscapes, or proximity to urban employment. Such activities have helped some rural communities survive the harsh reality that they no longer had an important agricultural purpose, other than facilitating the forced migration of farm families to the cities. However, most American rural communities continue to search for a new purpose for their existence.

If past trends affecting rural areas continue into the future, there will be little hope for revitalizing rural communities. But trends never continue, at least not indefinitely. A proposed list of the top twenty "great ideas in science" was reported in *Science* magazine in 1991, and scientists from around the world were invited to comment.[1] Among the top twenty were such ideas as the law of gravity, the relationship between electricity and magnetism, and the first and second laws of thermodynamics. The top twenty also included the proposition that "everything on the earth operates in cycles." Some scientists responding to the *Science* survey disagreed with the proposed theory of universal cycles, but many others left it on their list of the top twenty great ideas in science.[2] Based on the universal cycle theory, any observed trend is in fact just a phase of a cycle.

The theory of universal cycles implies that farms do not get either larger or smaller forever, but instead cycle between larger and smaller over time. If we think back over past centuries and around the globe, we can find examples when control of land became concentrated in the hands of a few, only to later become dispersed in control among the many. The most significant such occurrence in the United States may have been the development and later demise of plantation agriculture in the South. The most significant such occurrence in recent times probably took place in the former Soviet Union, where large collective farms were divided into smaller individual farming operations.

These cyclical turning points have been associated with major historical events. Today, large-scale industrial agriculture is coming under increasing environmental and social challenges in America and around the globe. The trend toward fewer and larger farms in the United States might also be a phase of a cycle that is nearing an end.

Historically, similar cycles have been observed in the spatial dispersion of people. Anthropological evidence indicates that people have concentrated in large cities in centuries past, but later, for a variety of reasons, they abandoned the cities and dispersed themselves across the countryside. Thus, there is reason to believe that migration from rural areas to U.S. cities during the twentieth century was simply a phase in a cycle rather than an unending trend. Many large city centers have already lost much of their previous population as people moved to the suburbs. A further migration back to rural areas, often labeled as urban sprawl, may be a logical continuation of the dispersion phase of this cycle. The most relevant question for rural communities might not be whether people will continue to abandon the cities and suburbs to resettle rural areas, but when and for what reasons. There is nothing in cycle theory dictating that people will return to the same rural areas they previously populated. They will need a reason to relocate to a particular place.

Alvin Toffler in his book *PowerShift* points out that many forecasters simply present unrelated trends as if they would continue indefinitely, without providing any insight regarding how the trends are interconnected or the forces likely to reverse them. He contends that the forces of industrialization have run their course and are now reversing.[3] He writes, "The most important economic development of our lifetime has been the rise of a new system of creating wealth, based no longer on muscle but on the mind."[4] He contends that "the conventional factors of production—land, labor, raw materials, and capital—become less important as knowledge is substituted for them. . . . Because it reduces the need for raw material, labor, time, space, and capital, knowledge becomes the central resource of the advanced economy."[5]

Knowledge-based production systems often embody enormous complexity in the form of simultaneous and dynamic linkages among a multitude of interrelated factors. Cognitive scientists have shown that humans can deal consciously and simultaneously with only a very small number of separate variables. Yet humans can perform enormously complex tasks, such as driving a car in heavy traffic, playing a tennis match, or carrying on a conversation, that baffle the most sophisticated computers and robots. People are capable of performing such tasks quite competently, if not routinely, by using their well-developed subconscious minds.

The subconscious human mind appears to be virtually unlimited in its capacity to cope with complexity. As organizational theorist Charles Keifer puts it, "When the switch is thrown subconsciously, you become a systems thinker thereafter. Reality is automatically seen systemically as well as linearly. Alternatives that are impossible to see linearly are surfaced by the subconscious as proposed solutions. Solutions that were outside of our 'feasible set' become part of our feasible set. 'Systemic' becomes a way of thinking and not just a problem solving methodology."[6] The subconscious human mind is capable of assimilating hundreds of feedback relationships simultaneously as it integrates detail and dynamic complexities together.[7] The human mind may be the only mechanism capable of dealing effectively with systems complexities that will characterize economic development in the future.

Peter Drucker, a noted business consultant, talks of the "postbusiness society" in his book *The New Realities.* He states, "The biggest shift—bigger by far than the changes in politics, government or economics—is the shift to the knowledge society. The social center of gravity has shifted to the knowledge worker. All developed countries are becoming post-business, knowledge societies. Looked at one way, this is the logical result of a long evolution in which we moved from working by the sweat of our brow and by muscle to industrial work and finally to knowledge work."[8]

Drucker contends that there is an important, fundamental difference between knowledge work and industrial work. Industrial work is fundamentally a mechanical process whereas the basic principle of knowledge work is biological. He relates this difference to determining the "right size" of organization required to perform a given task. "Greater performance in a mechanical system is obtained by scaling up. Greater power means greater output: bigger is better." But this principle doesn't hold for biological systems, where size is found to be consistent with function. "It would surely be counterproductive for a cockroach to be big, and equally counterproductive for the elephant to be small. As biologists are fond of saying, 'The rat knows everything it needs to know to be a successful rat.' Whether the rat is more intelligent than the human being is a stupid question; in what it takes to be a successful rat, the rat is way ahead of any other animal, including human beings."[9]

Differences in organizing principles may be critically important in determining the future size and organizational structure of economic enterprises and ultimately in determining their optimum geographic location. Other things being equal, the smallest effective size is best for enterprises based on information and knowledge work—"*bigger* will be *better* only if the task cannot be done otherwise."[10] Small enterprises can be located almost anywhere.

Robert Reich, U.S. secretary of labor in the Clinton administration, addresses future trends in the global economy in his book *The Work of Nations*.[11] He identifies three emerging broad categories of work corresponding to emerging competitive positions within the global economy: routine production service, in-person service, and symbolic-analytic service. He calls routine production workers the old foot soldiers of American capitalism in high-volume enterprises. This category includes low-and mid-level managers—foremen, line managers, clerical supervisor, and so forth—in addition to traditional blue-collar workers. Production workers typically work for large industrial organizations. These workers live primarily by the sweat of their brow, or their ability to follow directions and carry out orders rather than by using their minds.

In-person service, like production service, entails simple and repetitive tasks. The big difference is that these services must be provided person to person. This category includes retail sales workers, waiters and waitresses, janitors, cashiers, child-care workers, hairdressers, flight attendants, and security guards. Like routine production work, most in-person service workers are closely supervised and are required to have relatively little education. In-person services may be provided through a diversity of organizational structures, ranging from individual providers to large, franchised organizations. Unlike routine production work, individual personality can be a big plus, or minus, for in-person service workers.

Symbolic analysts are the mind workers in Reich's classification scheme. They include all the problem solvers, problem identifiers, and strategic brokers, such as scientists, design engineers, public relations executives, investment bankers, doctors, lawyers, real estate developers, and consultants of all types. They also include writers and editors, musicians, production designers, teachers, and even university professors. He points out that symbolic analysts often work alone or in small teams, which are frequently connected only informally and flexibly with larger organizations. Reich agrees with Toffler and

Drucker in suggesting that power and wealth in the future will be associated with symbolic-analytic service, by mind work, rather than by routine production or in-person services.

John Naisbitt and Patricia Aburdene in their book *Megatrends 2000* call the triumph of the individual the great unifying theme at the conclusion of this century.[12] They talk about greater acceptance of individual responsibility as new technologies extend the power of individuals. Their mind workers are called individual entrepreneurs. They point out that small-time entrepreneurs have seized multibillion-dollar markets from large, well-heeled businesses. In fact, during the 1980s, roughly two-thirds of all new nonfarm jobs were created by small businesses. An earlier National Science Foundation study showed that small businesses produced twenty-four times as many innovations per research dollar as did large businesses.[13]

Naisbitt and Aburdene contend that empowered individuals, while working alone or in small groups, will choose not to face the world alone but rather will seek community, which they define as the free association of individuals. Large business organizations, government bureaucracies, labor unions, and other collectives have provided hiding places for avoiders of responsibility. In a community there is no place to hide. Everyone knows who is contributing and who is not. In communities, individual differences are recognized and rewarded. The sense of community, all but destroyed by industrialism and collectivism, may well be restored by individuals empowered with knowledge. These people are looking for a place to be recognized, a place to belong, and not a place to hide.

Naisbitt and Aburdene talk of a new electronic heartland. They contend that a new breed of mind workers will reorganize the landscape of America. They will be linked by telephone, fax machines, Federal Express, and computers into information networks that span the globe. "Free to live almost anywhere, more and more individuals are deciding to live in small cities and towns and rural areas."[14] Many rural communities already are technologically linked to urban centers, and others will follow. The industrial revolution built the great cities of Europe, America, and Japan. But today's cities are based on technologies of a hundred years ago such as indoor plumbing, electric lighting, steel-frame buildings, elevators, subways, and telephones. Railroads and waterways made it easy to move raw materials and finished goods cheaply over long distances, but it was very expensive then to move people even short distances.

The cities have already lost much of their purpose as places for people to live. Multilane freeways and extended mass transit systems have allowed people to retreat to the suburbs by making it easier for them to get to and from work. But low-cost air travel has now reduced costs, in both time and money, of moving people over far greater distances. In addition, knowledge-based enterprises are far less dependent on movement of either raw materials or finished products. Most knowledge work can be delivered anywhere on the globe almost instantaneously at costs representing a very small fraction of its value. Mind workers are more independent of large organizations and thus require less frequent personal contact. For the first time in history, the link between a person's workplace and his or her home is being broken.

As Naisbitt and Aburdene point out, "In many ways, if cities did not exist, it now would not be necessary to invent them."[15] Drucker adds that the real estate boom and the associated new skyscrapers in big cities in the 1970s and 1980s were not signs of health, but instead were signals of the beginning of the end of the central city. "The city might become an information center rather than a center of work—a place from which information (news, data, music) radiates. It might resemble the medieval cathedral where the peasants from the surrounding countryside congregated once or twice a year at the great feast days; in between it stood empty except for the learned clerics and its cathedral school."[16]

People are abandoning the cities for the suburbs for quality-of-life reasons: lower crime rates, quality housing at a lower cost, and recreational opportunities. Many people are now free to abandon the suburbs for rural areas for quality-of-life reasons as well: more living space, a cleaner environment, prettier landscapes, and perhaps most importantly, a sense of community, a sense of belonging. The new challenge of rural economic development is to create places where mind workers can be productive and grow, where both immigrant and home-grown mind workers choose to stay and become part of the community.

Community economic development strategies are already undergoing significant changes to accommodate knowledge-based systems of economic development. As large companies and branch plants leave rural areas and move to other countries with cheaper labor costs, economic development professionals are beginning to concentrate on improving the quality rather than the quantity of jobs. The old strategies of industrial recruitment through building industrial parks by offering tax breaks is giving way to growth-from-within policies. The new strategies are in line with the business theories of Reich and others, investing in mind workers by encouraging entrepreneurs within the community to build small businesses and strengthen the local economy. Local buyer-supplier projects are encouraging rural people to plug the loss in dollars leaving their communities by replacing imports with locally produced goods and services.

However, most communities still seem to be lacking a clear vision of a new fundamental purpose for their existence. Many feel they can no longer depend on agriculture as the primary engine of rural economic development and are beginning to realize that industry recruitment is not a dependable replacement for most rural communities. There simply won't be enough American-based manufacturing operations in the future to go around. They see promotion of small-scale projects, such as niche markets, bed and breakfasts, and local festivals, as piecemeal, stopgap strategies with limited long-run potential for developing their communities.

Communities are seeking strategies for *sustainable* rural community development. They need development that is linked to local resources, that maintains the productivity of those resources, and that protects the local physical and social environment. However, sustainable development must also provide an acceptable level of economic returns and otherwise enhance the quality of life for those who live and work in the community. Development strategies that rely solely, or even primarily, on local natural resources alone are unlikely to fulfill these latter requirements. However, the obstacle of limited local resources can be overcome by those who have a clear vision of the new realities of economic development and a firm commitment to make their community a part of the new rural renaissance.

The linking of rural community development with local resources will become increasingly important, but far less limiting, in the knowledge-based era of economic development of the future. Robert Reich stresses that "the economy" is no longer local or even national in scope, but is truly global. Neither communities nor nations can depend on capturing and sustaining benefits from local capital, local industries, or even locally developed technologies in a global economy. Money, jobs, and technology can and will move freely to anywhere on the globe where they can be used to the greatest economic advantage. Thus, sustainable development must be linked to something that is not so easily moved.

Reich outlines two fundamental strategies for national economic development in a global economy. First, he advocates investment in infrastructure, including such things as roads, bridges, airports, and telecommunications access systems. Infrastructure has two important development dimensions. First, it facilitates productivity by making production easier and more efficient. Second, infrastructure is geographically fixed in the country where it is built. If producers want to use U.S. roads, bridges, airports, and communications access, they have to use them where they are, in the country that built them. Infrastructure in many respects serves the same function as geographically fixed natural resources in linking development of a specific location.

Reich's second and even more important development strategy is to invest in people. People who work with their minds will be the fundamental source of productivity in a knowledge-based era of the twenty-first century. If a nation is to be productive in the postindustrial economy, its people must be productive. Reich apparently depends heavily on national allegiance to keep productive people working in the nation that helped them develop their minds.

With one important added element, Reich's strategy for national economic development becomes a logical strategy for sustainable rural community development. Rural communities cannot depend on an allegiance of rural residents to their communities to keep productive people working in rural areas. People can and will move among communities within the United States during the rural renaissance if they do not have good reasons to do otherwise. Thus, communities must place a high priority on attracting new mind workers as well as creating places where their homegrown mind workers will choose to stay. The primary attraction of rural communities for current and future mind workers will be the promise of a desirable quality of life.

Quality of life is a product of the social, political, and economic terms by which people relate to each other, and of the terms by which they relate to the other elements of their physical and biological environment.[17] Quality of life is clearly affected by the quality of relationships among people and between people and their environment. Obviously, some observable factors such as employment, income, personal safety, economic security, and access to health care are important aspects of quality of life. However, quality of life is also affected by peoples' subjective judgments regarding such things as self-determination, freedom to participate, individual equity, freedom from discrimination, economic opportunity, coping ability, social acceptance, and treatment according to the accepted social principles of one's culture. Quality of life is about far more than just jobs and income.

The communities that survive and prosper during the rural renaissance will be culturally diverse. Diversity will be an important source of creativity, innovation, and productivity as well as an important aspect of quality of rural life. In rural communities, people will have an opportunity to know each other individually rather than simply to accept the stereotypes of cultural groups. Successful rural communities will consist of longtime rural residents, bright young people who choose to stay, returning rural residents, those born in urban areas, and those born in other countries. They will be Anglo-American, African American, Asian, Mexican, Canadian, European, South American, Caribbean, and East Indian, with a healthy mixture of other ethnic groups thrown in. Male and female, young and old, rich and poor, educated and less educated may be viewed differently, but all must be respected for their differences in the workplace and in the town halls of rural renaissance communities. Communities that fail to meet the challenges of the cultural renaissance will probably also fail to provide the quality of life necessary to participate in the economic renaissance.

Successful rural revitalization strategies for the future will be unique to each community. Routinized processes and recipes for success were a characteristic of the industrial era but not of the postindustrial era of knowledge-based development. However, the fundamental principles and concepts outlined above can provide guidance for those who have the vision of a rural renaissance and the determination to participate in this historic process. The following are a few of the more obvious elements of a successful rural revitalization strategy.

Invest in people. People are the basic source of productivity in a knowledge-based era of economic development. The "virtuous cycle" of education, increased innovation, increased investment, increased value, and higher wages offers an alternative to the vicious cycle of industrial recruitment, low wages, declining emphasis on education, declining communities, and the resulting downward spiral. The common practice of preparing the best and the brightest to leave rural areas must be reversed to meet both the cultural and economic needs of rural communities. Homegrown mind workers will value the quality of rural life that immigrants from urban areas will be seeking. High-quality, lifelong education will be equally critical to prepare people to succeed in the new, dynamic era of economic development.

Link development to local resources. Natural resources such as land, minerals, landscape, and climate must be utilized, at least initially, in the geographic locations where they exist.[18] Agriculture still has a key role to play in community development. Large-scale industrial agriculture operations provide little support for local communities. Sustainable agriculture, on the other hand, is a knowledge-based system of farming that depends on the productivity of local people. Sustainable farming is thinking farming. It requires an ability to translate observation into information, information into knowledge, knowledge into understanding, and understanding into wisdom. Certainly, sustainable farming involves hard work, but it is mostly a matter of thinking.

Sustainable agriculture is very much in harmony with a postindustrial paradigm of economic and human development. Sustainable farmers are *thinking workers* or *working thinkers*, as well as thoughtful, caring people. These new mind workers can multiply the value of agricultural products in rural areas and can replace many agricultural inputs currently brought in from elsewhere. Contrary to what some have suggested, rural communities need not abandon agriculture; they simply need to embrace this new kind of agriculture as a sustainable foundation for rural community development.

Invest in infrastructure. Good roads and access to airports will be important. However, modern telecommunications systems will be the key element in making rural areas competitive with urban and suburban areas in an information-driven, knowledge-based society. A national initiative to bring twenty-first-century communications systems to rural communities may be more important to rural areas today than were the rural free mail delivery and rural electrification programs of times past.

Invest in quality of life. Help people make the most of local climate, landscape, and recreational opportunities. Land use planning and zoning can make and keep quality spaces in rural communities, providing quality places for people to live. Make health care an investment in the future. Provide maternity wards and pediatricians, not just cardiac units and nursing homes. Make personal security and safety a top priority. These investments, as much as any single factor, will enhance the perception of rural communities as quality places to live.

Make a commitment to understanding, accepting, and valuing diversity. Quality of life depends on positive relationships among different types of people. Thinking, learning, behaving, and working alike were necessary for success in the industrial era of development. Thinking, learning, behaving, and working differently, but in harmony, will be the key to success in the knowledge-based era of development. Communities that fail to accept and value diversity among people are unlikely to succeed in embracing the different ideas that will be needed for success in the new knowledge-based era of development.

Share the vision. Rural communities must develop a shared vision of a positive future for rural America in general. The untapped demand for the quality-of-life attributes that rural communities have to offer is large and growing, but people have to be aware of where to find them. Productive people who desire a better quality of life may simply be locked into an old vision of rural communities as places of depression, decline, and decay rather than as places of new hope and inspiration.

The most important single step toward success may be for community members to develop a shared vision of what they want their particular community to be in the future. The vision of each person in the community will be different from the vision of others in many respects. However, the people of a community must search for and find some common elements among their different visions to form the nucleus of a shared vision of hope. Otherwise, the group is not really a community but rather a collection of people who happen to live in the same general area. A community that has found a shared vision of hope for the future has made its first critical step toward self-revitalization. To paraphrase prominent African American Jesse Jackson, if they can conceive it and believe it, they quite likely can achieve it. The future of rural America belongs to those who are willing to claim it.

¦ ¦ Presented at a symposium on Rural Community Development, sponsored by the National Center for Appropriate Technology Transfer for Rural Areas, Ferndale, Arkansas, May 7, 1993.

PART FOUR

The New American Farmer

12

Walking the Talk of Sustainable Agriculture

The sustainable agriculture movement is now at least a decade old. Some may be uncomfortable referring to sustainable agriculture as a movement. However, a social movement is nothing more than a sustained, organized effort by advocates of a common goal or purpose. Surely the organized efforts to develop a more sustainable agriculture have been advocated by enough people for long enough to qualify as a legitimate social movement.

The sustainable agriculture movement in the United States was validated when Congress approved a provision of the 1985 Farm Bill later dubbed "Low-Input Sustainable Agriculture" (LISA). The organized efforts leading to that action were begun at least a couple of years earlier. Some may contend that sustainable agriculture was just a continuation of the organic farming movement of the 1950s and 1960s. Others may argue that both the organic and sustainable farming movements simply continue earlier movements kept alive by farmers who refused to adopt the chemical-farming technologies that have dominated agriculture since World War II. However, the term sustainable agriculture did not come into widespread use until the late 1980s.

Sustainable agriculture represents a merging of three different streams of concern. Organic farmers and environmental groups were concerned about the impacts of agricultural chemicals on the natural environment and on human health. Some conventional farmers and agricultural groups were concerned about the impacts of rising input costs and falling prices on the agricultural economy. Small farmers and rural advocacy groups were concerned about the impacts of an industrial agriculture on farm families, rural communities, and society as a whole. These three groups joined forces to initiate the LISA program and remained united to defend it against attacks by agribusiness groups and their allies within the agricultural establishment. They saved the political identity of the movement in the 1990 Farm Bill by redefining and renaming LISA as the Sustainable Agriculture Research and Education, or SARE, program. Obviously, the SARE program is not synonymous with the sustainable agriculture movement; however, the persistence of the SARE program over the years, in the face of relentless efforts to disable or destroy it, bears testimony to the movement's continuing strength and durability.

Since the early 1990s, the sustainable agriculture movement has continued to grow from within, as it has picked up allies among other like-minded movements. The issues of economic globalization, corporate consolidation of the food system, confinement animal feeding operations, biotechnology, and other more general food safety, health, and nutrition issues have all helped to strengthen the sustainable agriculture movement. The movement now encompasses thousands, if not hundreds of thousands, of advocates and active proponents scattered across the continent and around the globe.

By the early 2000s, at least five annual sustainable agriculture conferences in the United States consistently were drawing more than a thousand people each year.[1] Sustainable agriculture conferences drawing 400 to 500 people were far from rare, and conferences drawing 100 to 200 people per year too numerous to count. Increasingly, farming conferences are planned in collaboration with citizen and consumer groups, or farmers are included in conferences sponsored by such groups. The sustainable agriculture movement is alive and well.

A lot of time and effort was spent in the early days of the movement trying to define sustainable agriculture. Some of the earlier questions concerning definition were genuine; sustainability was not a concept that easily fit accepted science-based classification schemes. Sustainability had undeniable social and ethical dimensions, which made many physical scientists both uncomfortable and skeptical. It was not a bottom-line economic issue, a fact that alienated the economic and agribusiness community. Many different definitions were suggested and many advocates proposed abandoning the word sustainable altogether; it was just too difficult to define and it seemed to alienate too many people. But the sustainable agriculture movement has persisted and its name has persisted with it.

Today, there is no longer any real lack of understanding concerning what sustainable agriculture means or what it requires—at least not among those who are willing to take the issue seriously. When someone today challenges an advocate to define what they mean by sustainable agriculture, almost invariably the challenger is simply trying to create confusion in the minds of others, to avoid being forced to address the very real questions of sustainability. They know intuitively that the answers to those questions will reveal the reality that conventional industrial farming systems quite simply are not sustainable.

Sustainability still doesn't have a simple little definition because it is not a simple little concept. But being difficult to define doesn't make the concept of sustainability any less important. Who is wise enough to provide simple little definitions for love, hope, faith, or even for profit? Yet few would argue that we can't deal with such things because we can't define them, or that they aren't important. People generally have a good understanding of the really big issues, such as love, hope, faith—and sustainability—even if they can't easily define them.

Nevertheless, it is certainly worth our continuing time and effort to try to find ways to communicate the concept of sustainability more effectively to those who have not yet linked the concepts of sustainability and agriculture. The most basic definition of sustainable agriculture is "an agriculture that will last"—an agriculture that is capable of maintaining its productivity and value to society, indefinitely. A sustainable agriculture must meet the needs of people of the present while leaving equal or better opportunities for those of the future. And in order to last, a sustainable agriculture must be ecologically sound, economically viable, and socially responsible. Lacking any one of the three, agriculture simply cannot maintain its productivity and value to society—it cannot last.

In somewhat different terms, the concept of sustainability applies the Golden Rule both within and across generations. We should take care of ourselves, if we are able, but also care for others as we would have them care for us, were we not able to care for ourselves. And we should care for those of future generations as we would have them care for us if we were of their generation and they were of ours.

Ben Franklin once suggested that philosophical and religious commandments such as the Golden Rule are not good for us just because they have been commanded of us, but are commanded of us because they are good for us. Caring for others is not a sacrifice but rather a privilege, because the positive relationships that result from our mutual concerns for each other are valuable, even essential, to a desirable quality of life. Stewardship of the earth for the benefit of future generations is not a sacrifice but a privilege, because it adds purpose and meaning, and thus quality, to our lives. Sustainability ultimately is about sustaining a desirable quality of life.

This is the rhetoric of sustainability. However, no matter how logical or persuasive the arguments, those who are unwilling to address the questions of sustainability will not be persuaded by rhetoric. Some will not believe an ecologically benign agriculture is necessary until they see and feel the impacts of ecological degradation for themselves. Some will not believe a socially just agriculture is necessary until their families are scattered or their communities are lost. Some will not believe an economically viable agriculture is particularly important until their farm faces an economic crisis. The only hope to reach such skeptics is to convince them that a sustainable agriculture will improve their quality of life, not some time in the future but here and now. Many won't believe that until they see the evidence for themselves.

Even the "true believers" in sustainability need tangible evidence that the prospects for creating a sustainable agriculture are real. Much of the optimism for a quick and easy transition to sustainability that bloomed early in the movement has faded with the passing of time and the continued resistance to change. However, the movement has not failed, as many of its opponents predicted. In fact, it has not even faltered. It is now obvious that the transition to sustainability will take more time and effort, but the questions remain as to how much time and effort it will take. Questions also remain concerning how we can convince others, and maybe even ourselves, that the movement eventually will succeed. Even the true believers need solid evidence that the rhetoric of sustainability can be transformed into a tangible reality.

Thankfully, thousands of farmers all across America and around the world are succeeding in transforming the rhetoric into reality; they are walking the talk of sustainable agriculture. These new farmers may label themselves organic, biodynamic, ecological, natural, holistic, practical, innovative, or nothing at all, but they are all pursuing the same basic purpose. They are on the frontier of a new and different kind of agriculture, an agriculture that is capable of meeting the needs of the present while leaving equal or better opportunities for those of the future—a sustainable agriculture.

While there is no blueprint for the new American farm, some basic characteristics are emerging.[2] First, these farmers see themselves as stewards of the earth. They are committed to caring for the land and protecting the natural environment. They value stewardship for ethical as well as economic reasons.

Second, these new farmers build relationships. They tend to

have more direct contact with their customers than do conventional farmers. They challenge the stereotype of the farmer as a fiercely independent competitor by freely sharing information and working cooperatively to do things they can't do as well alone. They value people for personal as well as economic reasons.

Finally, to these new farmers, farming is as much a way of life as it is a way to make a living. To them, the farm is a good place to live—a healthy environment, a good place to raise a family, and a good way to become a part of a caring community. Most new farmers are able to earn a decent income, but more importantly, they have a higher quality of life because they are living a life that they love.

The new sustainable farmer is a thinking farmer. Sustainable farmers must understand nature in order to work with nature, and they must understand people in order to build relationships with other farmers, neighbors, and customers. Agriculture has been characterized as the first step beyond hunting and gathering. Farming has been considered a low-skill, minimal-thinking occupation that almost anyone could do. Industrialization was said to be the next step beyond agrarianism—beyond agriculture. Higher-skilled factory work was considered a step up from farming. Certainly, sustainable farming involves some hard work, but success depends far more on thinking than on working. Sustainable farming is the mind work of the future, not the factory work of the past.

As with all true mind work, there are no recipes to guarantee success or sets of "best management practices" to insure against failure. These new farmers must fit their farming operation to the uniqueness of their farm—to their place within both natural and human communities. They must find their own market niche and develop their unique relationships with their particular customers. They must find a way of farming that fits their unique perception of a life of quality. That said, a couple of decades of experience with these new farmers, from all across

North America, has provided us with some insights into the general kinds of things that seem to be working for more than a few new farmers.

Grass-based livestock production seems to be among the most common of farming systems that are working well for sustainable farmers across the United States. Grass-based livestock operations include dairy, beef, poultry, pork, lamb, goats, and others. Free-range chickens, turkeys, and hogs fit into this general category as well, although for free-range animals, freedom to roam may be more important than access to grass. With grass-based operations, farmers increasingly are finding ways to make a good living with a fraction of the animal numbers they would need for a comparable way of life raising animals in confinement.

Grass-based dairies seem to offer the best economic opportunities with the least investment for those who have the necessary skills and temperament. By utilizing management-intensive grazing—sometimes referred to as planned grazing or rotation grazing—grass-based dairy farmers are able to reduce the high costs of purchased feed, equipment, fuel, repairs, and medication generally associated with confinement operations. Some milk producers cut costs further by milking only seasonally, taking maximum advantage of pastures by drying off all of their cows in midwinter. Grass-based dairy farmers are able to make more money, even while milking fewer cows and getting less milk per cow, because they are able to reduce costs through more intensive management.

The conventional wisdom in farming circles is that beginning farmers have to rely on off-farm income or a generous relative to get started in farming because of high capital requirements and low profit margins. However, the personal testimonials of a growing number of new dairy farmers are proving that it is possible to buy a farm and pay for it in a reasonable amount of time with a well-managed, grass-based dairy operation. The economic potential of grass-based dairies may become even greater as farmers find ways to translate the health benefits of products from cows with grass-based diets into market values. Switching from conventional to organic milk production also is a relatively easy step for the grass-based dairy farmer. Well-managed pastures require few chemical inputs. Less reliance on feed grains reduces costs of purchasing organic feed for grass-based organic dairies compared with conventional organic dairies. Whenever organic farmers are able to market milk directly to local customers, or even directly to local retailers, the value of their milk may be expressed in dollars per quart rather than dollars per hundredweight. In addition, on-farm milk processing—pasteurizing, homogenizing, and bottling—is becoming increasingly affordable, even for modest-sized dairy operations.

The potential for grass-based dairies is even greater for those with the skill and aptitude to turn milk into higher-valued specialty products such as cheese, yogurt, or ice cream. Some cheeses made from sheep's milk and goat's milk may sell for dollars per ounce. Of course, producing and marketing high-quality cheeses and other processed products from the milk of sheep, goats, or even cows requires highly specialized skills and often years of experience. In other words, the work can be very rewarding—personally, professionally, and economically—because it requires a lot of knowledge and creative thinking.

The potential for grass-based meat production from beef, sheep, and goats is similar to that of dairy, except that meat production typically requires more land and more livestock to generate a comparable amount of income. However, by marketing meat directly to local customers, meat producers can greatly increase the value of products sold per acre and per animal. Some grass-based meat producers increase both the efficiency of their intensively managed grazing system and the variety of their products through multispecies grazing of cattle, sheep, goats, and even poultry on the same farms. Also, the economic value from the greater health benefits of grass-fed meats may be easier to realize through direct sales to health-conscious customers. Economic limits are more a matter of being unwilling and unable to "think outside the box" than of the type of animal grown or products produced.

Organic production is another means of adding value to grass-fed meats. As with dairy, the transition to organic is relatively easier from grass-based production than from conventional meat production. When livestock producers sell directly to local customers, they also may receive premium prices for meats produced without hormones and antibiotics, raised under humane conditions, or given free range; all are highly compatible with grass-based systems. In some instances, doctors may recommend that their patients with allergies or potential sensitivities to antibiotics or hormones seek out producers who can supply meats without such additives.

Pastured poultry and free-range chickens and eggs are among the fastest growing of the new farm enterprises. I have never talked with a producer of pastured or free-range poultry who couldn't sell more birds or eggs than they were able to produce and process, at almost any price they choose to charge. On-farm processing has been the primary limitation, as government-inspected processing facilities for farm-raised poultry products have been very limited or nonexistent in most areas. First-time consumers of pasture-raised and free-range poultry and eggs become immediately aware that they are eating a fundamentally different product from the factory-produced poultry and eggs they have bought in the supermarket. The taste, texture, and color of free-range poultry and eggs are markedly different, in much the same way that vine-ripened, freshly picked tomatoes are different from gas-ripened, rock-hard tomatoes from the supermarket. Thus, issues of price and convenience become secondary.

Pork from hogs grown on pastures and in open facilities has much the same customer appeal as pastured and free-range poultry, with the same basic quality differences. Hogs by nature get a far larger proportion of their nutrition from grain or other concentrates than do ruminant animals. However, access to an outdoor environment, being able to forage for grass and insects and to root in the ground, affects the flavor of the meat. Also, the breeds of hogs supplying supermarket meats have been developed specifically for confinement production, where maximum pounds of saleable product at a maximum growth rate and minimum cost are the overriding objectives. Thus, flavor, texture, and substance have been sacrificed for the sake of economy. The traditional outdoor breeds of hogs tend to produce meat with more flavor and substance than do the confinement breeds. Pastured and free-range pork has an added advantage over beef, lamb, or poultry in that pork is highly marketable in processed as well as fresh forms. Cured pork and sausages provide excellent opportunities for enhancing the value, storability, and shelf life of pork products.

All grass-based and free-range animal products have the built-in advantage of being highly marketable to customers who are concerned about the social and ethical consequences of industrial food and farming systems. Grass-based systems are uniquely adapted to family farming operations because they rely on intensive management, meaning more management per acre and dollar invested, and thus smaller farms. Grass-based systems also offer a variety of opportunities for people with different skills and management abilities, and thus are well suited to family farms. Grass-based, free-range production systems are naturally humane environments in which to raise animals, since pastures are similar to the natural habitats of most farm animals. Certainly, animals can be made to suffer in such systems, but suffering is virtually unavoidable with factory systems of production. So, most well-managed grass-based and free-range systems result in products that can be marketed as raised under humane conditions on family farms.

Animal production systems need not be completely grass-based or free range to be legitimate family farms, to treat animals more humanely, or to minimize the negative environmental and social impacts typical of animal factories. For example, hoop-house hog production systems that utilize deep bedding and composting of solid waste to minimize environmental impacts are productive, economically viable alternatives to conventional hog factories. Many grass-based meat producers feed grains and other feed concentrates to make their products more acceptable to their customers, but animals may be fed grain while still on pasture or the confinement period may be minimized to maintain more natural systems of livestock production.

The key to success with sustainable livestock and poultry systems is to work with nature, giving animals their natural sources of nutrition in their most natural environment, to minimize costs of production and maximize product quality. In addition, to realize the full value of sustainable production, producers must develop and maintain relationships with customers who value the unique quality characteristics of sustainably produced products. Producers and customers alike must realize and appreciate the additional social and ethical benefits, the quality-of-life benefits, of supporting more ecologically sound and socially responsible systems of production.

In many areas, sustainably produced crops—such as grains, vegetables, and berries—seem to offer even greater potential for success than do livestock enterprises. Organic grain production has been the mainstay of the sustainable agriculture movement in the Upper Midwest and Great Plains of the United States and Canada. Large price premiums for organic soybeans, particularly for beans exported to Asian markets, sparked interest among many conventional corn-soybean producers in the Midwest. However, most soon discovered that growing crops organically requires a far greater understanding of soils, crops, pests, and people than does conventional crop production. Organic farming is not just farming without fertilizers and pesticides. True organic farmers must use alternative means of providing nutrients and managing pests; they must be thinking farmers.

Continued rapid growth in markets for organic grain has sparked interest among some of the larger commercial grain producers. New USDA standards for organic grain production provide farmers with a specific set of prohibited and allowed production practices, making it possible to produce certified organic crops without adopting an organic philosophy of farming. Many "industrial organic" growers simply substitute allowable organic inputs for prohibited chemical inputs rather than learn to work with nature. After they destroy the natural productivity of one piece of land, they simply move to another. It remains to be seen how long these "industrial organic" operations can continue, but they will almost certainly narrow the organic premiums and limit the near-term economic opportunities for philosophically organic producers.

Today, the most successful organic producers are moving away from marketing organic *commodities* and moving toward marketing organic *products*. Organic grain producers are finding ways to differentiate their grains from those of other producers—specifically from the "industrial organic" producers. They are finding niche markets for specific varieties and qualities of soybeans that the industrial producers are reluctant to grow because of low yields or stringent production and handling requirements. Some new farmers are growing long-neglected specialty grains, such as triticale, spelt, kamut, quinoa, or even popcorn. Some are cleaning, processing, and packaging their grains for direct sales to individual customers while soliciting and taking orders by mail, telephone, or Internet. Some are marketing their grains not only as organic or pesticide-free, but also as grown on family farms using socially responsible systems of production and processing.

Organic and locally grown vegetables, berries, and fruits are perhaps the most widely recognized of all successful sustainable farming systems. Retail markets for organic foods grew at a rate of more than 20 percent per year during the 1990s, with organically grown vegetables leading the way. A typical market garden relying on minimal equipment and family labor probably averages something like five acres in size and may return around $15,000 to $20,000 in returns to land, labor, and management.[3] Market gardens who rely on hired labor and field-scale equipment probably average around twenty-five acres and may return around $45,000 to $60,000 to the farmer's land, labor, and management. Most of the smaller producers market directly to their customers through farmers' markets, roadside stands, community-supported agriculture (CSA), or other direct marketing methods, and thus realize the full retail value of their products.

As with organic grains, strong market demand has sparked the interest of "industrial growers," and a few large organic corporations now control a large segment of the wholesale market for organic fresh produce. As with grains, independent family farmers have had to focus on direct marketing methods to maintain their economic viability. Many smaller producers have decided it is not worth the time, money, and effort for them to remain "certified organic" under the new USDA program. They will continue to farm organically and communicate directly with their customers concerning their farming methods, rather than rely on organic certification.

The number of farmers' markets and CSAs have grown so fast since the early 1990s that any reported statistic is likely to be woefully out of date. The important question for any given producer is not how many such markets exist, but rather the distance to the nearest farmers' market or the number of farmers' markets within a reasonable driving distance. Others may need to know the number of local CSA sites relative to the number the local community can support, the number of cars that pass their farms each day, or the number of people that would come to their farms if given a good reason to do so. In other words, the relevant question is what are my opportunities to market directly to customers in my community?

The primary advantage for sustainable vegetable, fruit,

and berry producers is their ability to choose plant genetics and growing methods for quality—flavor, nutrition, and variety—rather than for durability during harvest, transportation, storage, and display. The very best sweet corn, for example, must be eaten within a few hours of harvest, which is not possible with conventional systems of production and distribution. And the difference in flavor between vine-ripened tomatoes and supermarket tomatoes is legendary. In general, the greatest opportunities for sustainable production of vegetables tend to be for those located in urban fringe areas, or at least near reasonably sized population centers. Since two-thirds of all farms in the United States are located in "metropolitan" counties or in counties adjacent to "metropolitan" counties, direct marketing opportunities are quite common.

Fruits and some berries are less perishable than are most vegetables, and thus may be marketed to more-distant customers. Processing of fruits and berries into preserves, jams, juices, and so forth further increases marketing possibilities and widens the logical market area. However, product quality and distinctiveness and customer relations are no less important in distant markets than in local markets. Successful sustainable producers must offer products that are different, and better in the minds of their customers, than similar products available from elsewhere.

Organic markets also have been profitable alternatives for some fruit and berry producers, both domestic and export. Organic fruit production in particular seems to be more challenging than is organic grain or vegetable production and thus has been more difficult to industrialize. But organic certification or other types of eco-friendly or family-friendly labels cannot substitute for personal relationships in either distant or local markets.

The full variety of opportunities for sustainable farming is far too great to enumerate. Literally thousands of farmers all across North America and around the world are breaking away from the industrial system of farming and are finding new and better ways to farm and to live. As farmers seize opportunities to process and market cooperatively, the variety of opportunities will expand even further. But when sustainable farmers organize to process and market cooperatively, they must remain mindful that their advantage is in doing "something different"—something industrial food firms cannot do, or can't do as well. A small farmer-owned cooperative simply cannot compete with a multinational corporation using a mass-production, mass-distribution strategy. But the opportunities to do something different are virtually unlimited. And as more people become aware of the increased availability and variety of local foods, farmers' opportunities in localized, community-based food systems will explode. The important point for farmers is that far more market opportunities already exist than there are farmers who are willing and able to take advantage of them.

Obviously, all farmers do not succeed in their attempts to walk the talk of sustainability. Most who fail probably fail economically, but the lack of economic viability is often rooted in a lack of ecological integrity or a failure to provide an acceptable social quality of life.

Some new farmers come into farming from nonfarm backgrounds with unrealistic and idealistic expectations regarding the basic nature of farm life. Sustainable farmers are thinking farmers, but they are also working farmers. They can be thinking workers or working thinkers, but not thinkers who don't work or workers who don't think. This linking of working and thinking makes sustainable farms unique. Those who are willing to think but not work, or work but not think, or worse yet, neither think nor work, are destined to fail.

Some new farmers are physically unable to farm sustainably. A person lacking physical strength may be able to help manage a sustainable farming operation, but someone who knows and really cares about the farm must provide much of the physical labor. Physical strength can be built up through the exercise of hard work. So it is not necessary to be physically strong to begin farming, but it *is* necessary to become and remain strong to continue farming. There are numerous examples of farmers who were successful organic farmers in their early years but began to falter as they grew older. Some gave up and quit farming. Others, however, found ways to diversify into marketing and processing, turning more of the physical work over to other caring workers, and thus were able to continue. Over the longer run, each generation of farmers must find ways to bring younger people into their operations if their farms are to be truly sustainable. Sustainable farmers know that there is nothing wrong with hard work, it's actually good for us, but our physical ability to work hard inevitably diminishes with age.

Some new farmers fail because they are mentally unable to farm sustainably. It's not that sustainable farming is beyond their mental capacity, but many people simply cannot break away from the old mechanistic, industrial ways of thinking. Sustainable farming will never make sense to such people. They have a mindset that constantly reinforces their belief that the only way to do anything effectively is to specialize, standardize, and centralize decision making. They are incapable of "thinking outside the industrial box." For example, these farmers see organic production as just another set of "best management practices" and organic markets as just another way to "exploit misguided consumers." They may succeed in wringing a few more dollars of profits out of their farming operations in the short run, but their farms will never be sustainable.

Another fairly common reason for failure among so-called sustainable farmers is economic success. Some have become so successful financially that they have drifted back into old, industrial ways of thinking. After a while, they begin to make some "real money." They then begin thinking, if they just worked harder, borrowed more money, hired more workers, bought more equipment, if the operation was larger, they just might become wealthy. Soon they are working so hard that they don't have time to spend with their family or to enjoy life. They eventually lose all personal contact with their customers and no longer treat their employees like real people. Perhaps they are still making a lot of money, but nonetheless are miserable. The more fortunate sell their successful operation to some corporation that is only interested in making even more money. The less fortunate lose their health, their family, their farm, and sometimes their life, because they became "too successful."

This failure-through-success phenomenon is also common among farmers' cooperative ventures, particularly among the so-called new age cooperatives. Many such ventures fail because they are operated like farmer-owned *corporations* rather than farmer-owned *cooperatives*—that is, they don't do anything very different. Those few that succeed eventually become seen as direct competitors with their larger corporate counterparts. If they become sufficiently successful, their larger, more powerful corporate competitor may give them an option either to sell out for a profit or be driven out of business. Those who choose to sell may make a handsome return on their initial investment, but their cooperative venture will be over. Those who choose instead to take on the corporate world head to head are almost certain to fail.

The good news is that in spite of difficulties, frustrations, and occasional failures, more and more farmers are finding ways to succeed in walking the talk of sustainability. Thousands of new farmers all across North America and around the world are transforming the vision of sustainable agriculture into reality. These new farmers are learning how to work with nature to reduce their reliance on costly inputs that have polluted the natural environment and have squeezed the profits out of conventional farming. They have built relationships with other farmers, with their neighbors, and with their customers, helping to rebuild caring rural communities. They are quality-of-life farmers—those who realize that it is not a sacrifice to care for others and for the earth, but a privilege that adds quality to their own lives.

These new farmers are succeeding as grass-based livestock producers and as organic growers of grains, vegetables, fruits, and berries for local markets. They are succeeding by marketing directly to people who care about where their food was produced, how it was produced, and who produced it. They process and market together when it's to their advantage to do so, but they don't compete—they do something different.

These new farmers are building relationships of trust, integrity, honesty, and dependability with their customers, with their neighbors, and indirectly, with society. They value sustainable farming for ecological and social as well as economic reasons. They market to the growing numbers of potential customers who are willing to support their shared vision with their time and money. They are creating farming systems that can last, for the benefit of all people for all times. These new farmers are walking the talk of sustainable agriculture.

¦ ¦ Presented at the Annual Conference of the Nebraska Sustainable Agriculture Society, Aurora, Nebraska, February 22, 2003.

13

Survival Strategies for Small Farms

Over the past several decades, U.S. farms have grown larger in size and fewer in number. Farmers have substituted capital and off-farm technology for labor and management, making it possible for each farmer to farm more acres—utilizing more hired labor, equipment, and facilities—thus leading to fewer farmers and larger farms. Today, the large farms that have survived consolidation thus far are increasingly coming under the control of gigantic multinational agribusiness corporations, through comprehensive contractual arrangements, thus continuing the trend toward fewer and larger farming operations.

For decades, farmers have been told that they will have to either get bigger or get out of farming—that small family farms were "a thing of the past." Virtually all government farm programs, including federal loan programs, have unwittingly supported this trend toward fewer and larger farms in their preoccupation with helping farmers increase their productivity. The few government programs targeted specifically to small farms, such as small farm loans, beginning farmers, direct marketing, and 1890 land-grant extension programs, are considered by many as doing little more than prolonging the agony of a dying way of life.

Understandably, it may be hard for those who work with farmers to get very enthusiastic about promoting a way of farming that is supposedly doomed to extinction. But many small family farms have survived, and at times have even prospered. Somehow, small farmers have found ways to survive and succeed, in spite of the misguided government programs and outdated public perceptions that have created obstacles to their survival and success.

Large industrial farming operations have succeeded, at least in part, because they have been the recipients of huge government subsidies, not only through direct government payments but also through government subsidies for farm credit, public research and extension services, and export promotion. The traditional mid-sized, full-time family farm has been pushed to the brink of extinction in America because they have neither the political clout of the large agribusiness enterprises nor the resilience, resistance, or regenerative capacities of small farms. While undoubtedly well intentioned, government farm programs have been major contributors to the demise of full-time family farms.

Small farms, on the other hand, have succeeded in the past and can continue to succeed in the future, even without a "level playing field" in terms of government programs or public understanding. But the odds of success for any individual farmer could be considerably enhanced if current misperceptions concerning the imminent demise of small farms were replaced with the new realities of small farm opportunities.

Outdated perceptions concerning small farms are deeply rooted in the institutional culture of USDA, the land-grant universities, and other public agencies, as well as in the minds of the general public. In addition, powerful economic and political interests oppose any change in the public agenda that would better serve the needs of those currently without economic or political power, which includes those on small farms. These groups work hard to reinforce the current misperceptions in order to protect their own special interests—to keep their place at the public trough. Only when these outdated perceptions are forced to confront today's realities will the full measure of opportunities for small farms be realized.

One such outdated perception is that small farms are not a significant part of agriculture. Agricultural programs for the past several decades have been driven by concerns for production rather than people. The underlying assumption was that the public would benefit most by focusing on improving the efficiency of farming, ultimately bringing down the cost of food and fiber to consumers. This focus on efficiency has been the root source of the trend toward larger, more specialized farming operations. As large farms accounted for an increasing share of total production, the remaining small farmers had a diminishing impact on overall food supplies and prices, and thus became less important to the agricultural institutions.

Today, USDA and the land-grant universities are promoting high-tech and biotech production methods for the same reason: cheap food for consumers. The natural environment is viewed as a constraint, not an asset, and it doesn't seem to matter whether there will be any farmers left in this country in a decade or two, or whether rural communities survive or die. They see their public mandate as ensuring that agriculture is as efficient as possible so consumers will have an abundant supply of food at minimum cost. Small farmers are simply not relevant to that mission.

Another perception is that small farms are not real farms. Most small farms are part-time farms; many are nothing more than rural residences with a garden or a few head of livestock. Others are considered to be strictly hobby farms, not intended to earn an income from farming. In addition, some farmers are urban residents who own land in the country. When the 1997 census definition was changed to include farmers whose entire farms were in the Conservation Reserve Program and other nonfarming farmers, the ranks of small farmers was expanded considerably. While living in the country or owning land may be important to such people, there is little if any income derived from their actual farming operation.

The current perception is that it is simply not realistic for farmers to depend on a small farm for a significant part of their economic living. Many see no way that a farm with gross sales of less than $50,000 a year can be a serious commercial operation. Farmers' net incomes generally run about 15 to 20 percent of gross sales, even on well-managed small farms, and $7,500 to $10,000 a year certainly won't support a family. So the USDA categorizes such farms as "noncommercial." Even farms with gross sales of up to $100,000 per year would be expected to net only $15,000 to $20,000 in income to support a family.

Minority farmers are seen as even less important because they make up a very small percentage of those farmers who have little if any chance to survive and succeed in farming. A common perception is that lack of income for those few who are actually trying to make a living on small farms is a public welfare issue, not an agricultural issue. Government farm programs were never intended to be rural welfare programs. Those farms grossing between $100,000 and $250,000 are actually mid-sized farms, not small farms. These farms are thought to have a chance of surviving, but only if they get larger. Smaller farms are given little chance of ever becoming real farms.

Another common perception is that technologies developed for larger commercial farming operations are equally useful on small farms—that agricultural research and technology transfer programs are scale-neutral. After all, the only way for a small farm to survive and succeed is to get larger—to grow into efficient technologies. And those who can afford to farm as a hobby surely want access to the best technology available. It makes no difference to a cow whether she is in a herd of ten or ten thousand; her needs are still are the same. It makes no difference to a corn plant whether it is in a field of ten or ten thousand acres; its needs are still the same. If the scale of technology doesn't matter to the animal or plant, it is of no consequence to the farmer, according to the apologists for scale-neutral technology.

A related perception is that the needs of small farmers are being met by existing government policies and services for agriculture. If anything, small farmers get more than their share of government services, so the argument goes. After all, some programs are designed specifically to meet the needs of smaller family farms, and maximum government payments per farm are limited for some commodity-based programs. Larger farms actually should get a greater portion, they argue, because program benefits have to be focused on those who produce the bulk of agricultural commodities if those benefits are to have the greatest impact on production and prices.

Those who complain about inadequate attention to small farms in public research and education are seen as living in the past, when small farms were actually economically viable. Progress in the agricultural economy by necessity has simply left small farmers behind. There is little government could do to roll back, or even to slow, the technological advances supporting large-scale agriculture, even if it could justify doing so. Small farm advocates are viewed as simply being out of touch with reality. Small farmers are citizens and thus worthy of attention and indulgence by those in public institutions, but USDA can do little for them that is consistent with the perceived USDA mandate of increasing productivity. To USDA and land-grant universities, the small farm issue is a public relations issue, not a legitimate agricultural issue.

However, the realities of small farms today are very different from these perceptions. In some cases, prevailing perceptions are simply out of date, but in others they are simply wrongheaded.

The reality is that small farms are a significant and important part of American agriculture.[1] While a focus on agricultural productivity may have been legitimate in the past, there is no longer any significant societal benefit to be gained from continued public programs designed to enhance the productivity of agriculture. First, American consumers no longer spend 40 to 50 percent of their income for food, as they did when USDA and the land-grant universities were established. Instead, today Americans spend a little more than 10 percent of their income, a dime of each dollar, for food. In addition, the farmer only gets to keep about 10 percent of each dime consumers spend for food; the rest goes for purchased inputs and marketing services. Even if farming were perfectly efficient, if the farmer kept nothing, consumers would save only a penny from each dime spent at the grocery store—a penny of each dollar of disposable income. Government farm programs simply cannot make food much cheaper. And it certainly no longer makes sense to try to make food cheaper by making farms bigger.

Furthermore, government can no longer justify subsidizing those who produce the bulk of the nation's food—the 17 percent or so of the producers who account for some 80 percent of total production. These are the large agricultural enterprises operated by people with higher incomes and far greater wealth than the average American taxpayer. While relatively few of these are listed as nonfarm corporations, many are contract producers for large, multinational agribusiness corporations. The corporations make most of the profits from such operations, with most of what's left going to those who own the land or finance the production facilities. The "farmers" in such operations are little more than contract laborers and landlords, receiving little more than minimum wages and rents.

The original mission of USDA and the land-grant universities was to support agriculture with public funds because agriculture was fundamentally different from industry. Farmers confronted different economic forces than those confronted by industry, and farmers' operations weren't large enough to finance their own research and technology development. These were the basic justifications for public support for farming. However, agricultural commodity production today is simply another industry, and giant agribusinesses neither need nor deserve government protection; instead, the public needs protection from them. Agribusiness firms are large enough to do their own research and development. There is no longer any justification for using taxpayers' dollars to subsidize industrial agriculture.

The primary public issues confronting agriculture today are

ecological and social in nature. We must have people on the land who care about the land if we expect our land and other natural resources to be cared for and be capable of sustaining American society in the future. Farmers—the *people* who farm—are still the backbone of many rural communities and are the primary keepers of America's rural culture. Farmers—the *people* who farm—are more important to American agriculture today than are the quantity and price of agricultural production. The reality is that most of the *people* who farm in America are on small farms.

The reality is that most real farmers are small farmers. Admittedly, some of those census entities counted as farms are hobby farmers and rural residences. But many are not. The Census of Agriculture survey asks farmers to state their "primary occupation"—the occupation at which they spend more than half their working hours. Recent surveys also distinguish between active farmers and retired farmers. Small farmers are more likely than large farmers to have some occupation other than farming and are more likely to be retired. But even when considering only those whose primary occupation is farming and who are not retired, more than half of all farmers would easily be classified as small farmers. Well over half of the primary-occupation farmers have less than $100,000 in annual gross sales. Nearly half have gross sales of less than $50,000 per year—classified by USDA as noncommercial farms. Should one's primary occupation be called "noncommercial?" Most *real* farmers are small farmers.

The reality is that many of these small farmers do earn a living, or at least a significant portion of their living, on such small farms. These successful small farmers pursue a fundamentally different approach to farming than do large farmers. They are low-input farmers; they reduce their reliance on purchased inputs and borrowed capital by substituting management of their internal resources—land and labor. In general, they substitute intensive management for purchased inputs and family labor for borrowed capital. They also focus on creating value, as well as reducing costs. They are niche marketers, many of whom market directly to local customers and gain a significant share of the 80 percent of food value that usually goes to middlemen. Because of these things, many small farmers earn far greater income per dollar of sales than do conventional large farmers. A farm with $50,000 in gross sales, for example, may well contribute $25,000 or more to support a family.

Second, many small farmers live simply. This does not mean that they live in poverty, but it does mean that their economic standard of living may not be as high as that of their urban neighbors. To them, the primary product of their farm is a desirable quality of life. The farm provides them with a home, much of their food, a place for raising a family, an aesthetically pleasing place for recreation and relaxation, and a place for learning and teaching, as well as a place to work. Many of these smaller farms are not obligated to report a net income from farming because many of the costs of living on a small farm qualify as "farm costs" for income tax purposes. Many of these farms, particularly if they are part-time farms, need not earn an income from farming. The nonmarket value of the farm to the family is sufficient to justify the farm being the primary occupation of at least one adult family member.

In general, the most successful small farms are following many of the philosophies and practices of sustainable agriculture. They are balancing the ecological, economic, and social aspects of their farming operations to support a desirable quality of life for themselves, their families, and their communities. They are exploiting neither their natural resources nor other people in their pursuit of profits. By doing what makes sense to them, economically, socially, and ethically, they are building a more sustainable system of food and fiber production for the future. The reality is that small farmers are the real farmers of the future.

The reality also is that the technologies developed for larger, commercial farming operations are not appropriate for small farms. Successful small farms must be management intensive—they must earn more returns per acre, per dollar invested, per dollar of production. The higher net returns on intensively managed farms come from the efficiency with which various practices, methods, and enterprises are integrated together, not necessarily from the efficiency of each individual practice, method, or enterprise. A cow plays a very different role in a complex integrated farming system than in a specialized beef or dairy herd. Corn plays a very different role in a complex integrated farming system than in a specialized corn or row-crop farm.

Small farmers need research and technology that will enhance their human capacity to manage things—to understand, to think, to learn to integrate things more effectively. They don't need technologies that require them to follow specific practices and procedures or restrict their options, in effect restricting their ability to manage.

The reality is that technologies are not scale-neutral. Agricultural technologies of the past have provided means of simplifying and controlling production processes, and thus have constrained the farmer's ability to manage. In fact, technology development and transfer programs from USDA and land-grant universities have fueled the industrialization of agriculture and have forced farmers to move toward ever larger and increasingly specialized farming operations. Small farmers need new ways to earn a better income with less land and less capital, not new ways to manage more land and more capital.

The reality is that existing government policy and services are not meeting the needs of small farmers. Government policies have not been focused on "saving the family farm," as the politicians have claimed, but instead on enhancing efficiency of agricultural production. Now that subsidizing greater productivity has no social justification, agricultural programs should be refocused on the needs of people, not production. The public now has a greater stake in farmers protecting the natural environment and supporting viable rural communities than in providing unnecessary incentives for giant agribusiness corporations to grow still larger. There is no more justification for subsidizing industrial agriculture than for industrial textiles, construction, steel, chemicals, or computers. The giant corporations have a profit incentive to develop all these industries, including agriculture, without government subsidies. The new role of government should be to serve the public good by protecting society and the natural resource base from corporate exploitation.

In all public programs, including those for agriculture, each person should be afforded equal worth and thus given equal consideration. All persons have an equal claim to public goods and services, no matter how rich or how poor they may be, no matter large or how small their economic contribution, no matter what their race or ethnicity. In addition, all persons are entitled to certain fundamental rights and privileges and have certain fundamental responsibilities, no matter how much or how little political power they possess. This is the nature of a democracy. In the private sector, we vote with dollars—the more dollars, the more votes. But in the public sector, everyone is equal—those with more dollars still get just one vote. If USDA and the land-grant universities truly functioned as public institutions, each farmer would be given the same importance and attention, no matter how small. If these institutions functioned as true public institutions, small farm programs would receive well over 90 percent of all program benefits because small farmers make up over 90 percent of all farmers. At a minimum, small farms deserve over half of all public resources, because over half of those who consider farming their primary occupation live and work on small farms.

Small farm advocates are in touch with the reality of today and are not living in the past. But perhaps more importantly, they are looking to the future. They are not opposed to new technology; they simply want technology that is consistent with long-run sustainability as well as short-run profitability. The future of human civilization depends not only on food but also on a healthy environment and a civilized human society. There can be no greater public priorities. No better means of sustaining human life on earth exists than that of having people on the land who are intellectually capable, socially dedicated, and ethically committed to meeting the needs of the present as well as the future through farming. No better investment of public dollars exists than investments in keeping the land in the hands of ecologically and socially responsible small farmers.

The reality is that small farms have managed to succeed in spite of persistent public misperceptions. And in these successes of the past are the keys to the survival and success in the future for those who choose to operate small farms.

Successful small farmers must think for themselves. Very few people really understand how to make a good living on a small farm. Most of the so-called experts have been taught how to help small farmers manage like large farmers, not how to manage their small farm. Most so-called best management practices and farm business strategies are designed to tell farmers how to make more money by managing more land and capital, rather than how to make more money with less land and less capital by managing better.

The best source of outside advice for small farmers is other small farmers, or those who have learned from other small farmers. But every successful small farm will be fundamentally different from every other small farm. So ultimately, small farmers must think for themselves and make their own decisions.

Successful small farmers must think like a small farmer—not like larger farmers who don't have enough land or capital. I am frequently asked how I define a "small farm." My typical answer has become, "The difference between large and small is in farmers' heads, not in how many acres they farm or in the size of their bank account." A farmer who is farming forty acres but feels that he or she needs to have more land to make a decent living is a *large* farmer, that is, thinks like a large farmer. A farmer who is farming three thousand acres but feels that he or she needs to find some way to make a better living while farming less land, is a small farmer, that is, thinks like a small farmer. A successful small farmer is one who finds ways to do more with less, while a large farmer always needs more. Successful small farmers are those who think small.

Successful small farmers know when they have "enough." The fatal flaw in conventional American farming was that farmers never knew when they had enough. They always wanted more—more land, more livestock, or more money. So they succeeded only in driving each other out of business, as each had to have the other's land in order to succeed. Eventually, however, even the survivors will be forced out of business by the large corporations. A farmer that defines success as "having more land, more livestock, or more money" will never be successful. He or she will never have enough. A farmer who defines success as enough to live a good life has a far better chance for success. It doesn't take a large farm to make a good life, but it does take knowing how much is enough.

Successful small farmers will be "quality of life" farmers. Quality of life is not something you acquire or accomplish; it is something you are—a state of being. It is not a product but a process. You don't possess a life of quality; you live a life of quality. Our quality of life is not determined solely by our income or wealth, although we do have economic needs that must be met. The quality of our life also depends on the quality of our relationships with other people, within our families and communities. A life of quality is a life of purpose and meaning—a life lived according to one's moral and ethical principles. When our pursuit of income and wealth degrades our relationships with others, it diminishes rather than enhances our quality of life. When our pursuit of income and wealth causes us to compromise our moral and ethical principles, it diminishes rather than enhances our quality of life. We will never find happiness if we spend all our time and energy pursuing material success.

The future of American agriculture may well depend on a reorientation of government programs to focus on the well-being of people rather than the efficiency of production. An industrial agriculture is inherently tied to production of basic, low-value agriculture commodities in which the United States is least likely to maintain a global competitive advantage. The integration of farmers into the new value-adding corporate food chains does not free the farmer from being a producer of low-value raw material to which someone else will add value and reap the profit. It has been estimated that more than 30 percent of all U.S. agricultural production is already produced under corporate contracts.[2] Ultimately the "value-adding" sectors of American agriculture will be completely controlled by large corporations, not by farmers or even groups of farmers.

The future of farming in America is at risk. We must shift the emphasis on farming from greater economic efficiency and global competitiveness to greater environmental integrity and national food security for the benefit of people, not corporations. America's small farmers will continue to plant another crop or keep cattle on pastures for as long as they can scrape together enough money to do it. It doesn't really matter all that much to most if they could make more money doing something else or if they could make more money farming in another country. As long as they can get enough money to buy seed and fertilizer, they are going to grow something, even if they are free to quit farming if they choose to do so. They have roots in their communities and they are not going to leave their families; they are committed to farming in America.

Multinational corporations have no such commitment to America and certainly not to farming in America. Corporations will help their contract growers get loans to buy buildings and equipment, but they will abandon those growers if the contractual arrangement becomes unprofitable for the corporation.

These corporations will invest their capital and apply production technologies wherever on the globe they can minimize costs and maximize profits. Those who insist on such complete economic rationality in decision making are supporting the ultimate abandonment of American agriculture.

American taxpayers should not be asked to support government programs that subsidize the exploitation of the natural environment for the sake of cheap food. American taxpayers should not be asked to support programs that subsidize the exploitation of people—neither contract growers in the United States nor peasant farm workers in other countries. Government programs should be restricted to independently owned and operated family farms—to farmers who will pledge their commitment to protecting the environment and to treating all living things, including other people, with dignity and respect. Government programs should not support the global "race to the bottom" to see who can minimize economic costs by imposing environmental and social costs on the rest of society.

Sustainable agriculture offers a viable alternative to corporate industrialization. A sustainable agriculture seeks to work in harmony with nature—to restore, renew, regenerate, and sustain the productivity of the natural environment. A truly sustainable agriculture would empower people to enhance their social and ethical quality of life, thus eliminating their need for continual economic exploitation of the earth and of each other. A sustainable agriculture is based on the belief that there are fundamental laws of nature, including human nature, that we humans violate only at our own peril. And a truly sustainable agriculture is easier to achieve on small farms.

Obviously, those who administer federal farm programs are not in a position to rewrite the policies and rules they must follow in carrying out their work. However, in many cases, the means by which government programs are administered are just as important as the authorizing legislation. An agency worker who is preconditioned to view a program application made by a large-scale specialized farming operation as a good prospect can almost always find a legal reason to approve the application. And an agency worker who is preconditioned to view the application of a small, diversified farming operation as a poor prospect can just as easily find a legal reason to deny the application. Perceptions of administrators can be as important as policy.

Those who administer federal farm programs need to be made aware that the fundamental nature of agriculture in America is changing. Today, large-scale specialized producers of basic agricultural commodities have the poorest prospects for the future. They will not have the access to the technology or markets they will need to compete with corporate agribusiness. Those who survive by becoming corporate contract producers will be forced to assume far more economic and environmental risks than can be justified by the meager rewards of contract farming. The smaller, diversified, part-time farmers have far better prospects for the future than do the large, specialized producers. They are not competing with corporate agriculture; they are doing something fundamentally different.

However, many potentially successful small farmers quite likely will be denied legitimate opportunities because of outdated perceptions on the part of those who decide who gets access to government programs. Federal farm programs have important roles to play in helping to shape the future of American agriculture. Perhaps they cannot determine who succeeds and fails but they can affect the balance of opportunities. Government programs can continue to support the trends of the past or they can help show the way toward a better future for American agriculture.

Polluted water, fouled air, mistreated animals, and oppressed workers are symptoms of an agriculture that has been forced to focus too narrowly on the economic bottom line. Chronic farm financial crises, declining exports, and loss of food security are symptoms of an agriculture that is being shaped by a global, corporate economy. Broken farm families, decaying rural communities, and "worn out" farms are all symptoms of farmers trying to be successful by getting bigger so they can be more productive and more globally competitive. Those who have created these problems are not bad people; they have simply believed the outdated perception that farmers had to either get bigger or get out of farming.

The future of farming in America is a way of farming that balances ecological integrity and social responsibility with economic viability. The future of farming depends on farmers—real people—who balance the personal, the interpersonal, and the moral and ethical dimensions of their lives. Bigger farms haven't resulted in better lives, not for farmers and not for society in general. The future of farming in America, the future of America, depends on small farms.

⁞ Presented at a Farm Loan Program Training Meeting, USDA Farm Service Agency, New Orleans, Louisiana, November 27–30, 2001.

14

Marketing in the Niches for Sustainability

Modern industrial farming methods are widely heralded as the world's most productive, but are they sustainable? Admittedly, U.S. food consumers spend little more 10 percent of their disposable income for food, but what are the environmental costs of producing the world's cheapest food? Industrialization has allowed fewer farmers to provide a growing population with more food and fiber at a lower cost, but what are the social costs of displaced farm families and dying rural communities? Industrial technologies have allowed U.S. farmers to reduce costs and increase production, but how many of the remaining two million farms are economically viable over time? The productivity of agriculture cannot be sustainable over time if the only profitable farms are those that deplete the environment and degrade the quality of life of farmers, rural residents, and society as a whole. On these grounds, U.S. agriculture most certainly should be questioned, if not indicted, for its lack of sustainability.[1]

The answer to the questions of sustainable farming will not be found by farming more efficiently—doing things right—but instead by farming more effectively—doing the right things. Those who have been most successful in pursuing sustainability almost always tell of beginning their quest by rethinking farming from the ground up. They talk of changing their heads first; changes in their farming then followed. Farming sustainably depends on doing better things rather than on doing the same things better.

Farming sustainably will require more intensive management of the farm's natural and human resources. Most of the early emphasis in sustainable agriculture was on production management. USDA's name for the initial sustainable agriculture program was "low-input sustainable agriculture," or LISA, suggesting less reliance on purchased or external farm inputs such as fertilizers, pesticides, hired labor, rented land, and borrowed money. However, if low-input farms are to remain productive and profitable, they must substitute something else for the purchased inputs they are eliminating. They must use some other means of maintaining soil fertility, managing pests, getting the work done, and financing the operation. The "something else" that makes low-input farms profitable is the farmers' own resource base—owned land, operator and family labor, and equity capital. Low-input farmers manage these resources more intensely by utilizing crop and pasture rotations, diversifying crops and livestock systems to maintain fertility and manage pests, using family labor to do the work and to reduce equipment costs, and keeping investments low to reduce reliance on borrowed money.

The logic of low-input farming can be applied to the process of marketing as well as production. In fact, marketing begins with decisions regarding what to produce. Producing higher-valued products rather than mass-produced commodities is the logical place to begin generating more net income without increasing land or capital. The focus on higher value should be on qualities inherent within the natural product, such as taste, freshness, or nutritive value rather than cosmetic qualities that can be enhanced more efficiently through use of commercial inputs. Sustainable farmers will be more successful marketing higher-valued products by also targeting individual customers and small groups of customers that are least well served by the existing industrial food markets.

The dramatic growth in organic foods during the 1990s has resulted from farmers producing for a specific market that was not well served by the large supermarket chains. Likewise, the growth in "green" markets for all sorts of "natural" products has arisen from growing public concerns about the environmental impacts of industrial production methods. Customers who are not well served by the current industrial system will reward those who are willing and able to respond to their unique preferences with products that meet their specific needs. Products can be tailored to meet the needs of specific customer groups by a variety of means, beyond simply deciding what to produce. Unique ways of processing and packaging may differentiate one farmer's produce from others in ways that better meet the needs of a particular group of customers. Home delivery, convenient pickup, or even inconvenient but authentic visits to farms to pick up food products, may distinguish one farmer's products from another. Providing fresh, local products before or after the normal local growing season—by using innovative means of production, processing, and storing—may provide a unique advantage for others. The key concept is not to compete, but to produce, process, and deliver something unique.

This strategy for targeting products to unique markets—targeting relatively small groups of customers rather than mass marketing—is commonly called "niche" marketing. Niche marketing may well be the key to sustainable farming, at least to economic sustainability. First, according to USDA statistics, about 80 cents of each dollar spent for food is attributable to processing, transportation, storage, advertising, and other "value-added" activities that typically occur beyond the farm gate.[2] Management-intensive production strategies must focus on economic advantages by squeezing more out of the farmer's 20 cents of the consumer's food dollar by reducing reliance on commercial inputs. Management-intensive marketing strategies can focus on the 80-cent marketing costs, which typically go to marketing firms, to enhance the economic viability of ecologically sound and socially responsible farms.

Cutting out the middleman gains nothing for farmers unless they can do the marketing job more effectively and efficiently than commercial food marketing firms. The primary advantage of mass marketing is that processing, transportation, storage, packaging, and advertising can typically be done at a lower cost on a large scale. There is no way an individual farmer can match the costs of a large agribusiness firm in carrying out the same marketing functions for the same products.

However, a high degree of uniformity and consistency among products is required for commercial middlemen to achieve the cost savings of mass marketing. Processors need to start with large quantities of uniform commodities or raw materials to facilitate efficient processing. They must turn these raw materials into large batches of uniform products to facilitate efficient transportation, storage, advertising, and merchandizing to masses of consumers. Differences among final products generally are limited to product presentation, such as cosmetic appearance, displays, and packaging, and superficial differentiation, such as color, shape, and size. In the case of broiler chickens, for example, there may be dozens of package sizes, varieties of cuts, and degrees of preparation, but they all must come from the same generic chicken in order to achieve the economies of mass marketing.

The strength of niche marketing arises from the weakness of mass marketing. Niche markets focus on supplying relatively small quantities of unique products with differences that go beyond presentation and cosmetics. Organic vegetables are a prime example of successful niche markets. The organic difference goes all the way back to the land and the farming methods under which organic crops are grown. These differences are preserved through the marketing process to ensure that organic consumers receive a product that reflects their values. Rather than minimizing costs, niche marketing focuses on maximizing value to the final customer.

Different people value things differently. Niche marketing focuses on getting the right product to the right person at the right place at the right time. Niche marketing responds to the unique preferences and individual needs of specific customers, giving them what they want rather than trying to convince them to accept what everyone else seems to want. The more unique the product, the higher the potential premium in value over similar products available in mass markets. Niche markets focus on value rather than cost, allowing niche marketers to avoid head-to-head price competition with mass marketers.

Niche marketing gives farmers an opportunity to get off the industrial treadmill of the past and to sustain the profitability of farming. It creates economic value by matching the unique resources of farmers with the unique wants and needs of consumers—a proposition that is validated by economic theory. All economic value or utility arises from four fundamental sources: form, place, time, and person or possession. In order to know the value of anything, we must first know its physical form (What is it?), its geographic location (Where is it?), its time of availability (When can I get it?), and finally, the people involved (Who has it and who wants it?). Only when we know the answer to all four questions can we know the economic value of anything.

During the industrial era, our attention has been focused on the first three dimensions of value: form, place, and time. The fundamental advantage of industrialization is that it greatly reduced the costs of changing the physical form of things, through mass production, processing, and manufacturing; changing the place of things, through transportation and distribution; and changing the time of things, through packaging and storage. So the focus of industrialization has been on the things for which industrialization provided clear advantages.

However, value is also associated with the personal dimension (Who has it and who wants it?). Different people have different abilities as producers. In economics, it matters who is doing the producing and who is offering something for sale. One person may be able and willing to produce and offer a superior product at a lower cost than can another. In economics, it also matters who is doing the buying. The same thing at the same place at the same time may have a greater value to one person than to another. It matters who wants, or doesn't want, what is offered for sale. But industrial systems simply can't create benefits for people as individuals, at least not very efficiently. Industrial systems must produce for people as masses in order to achieve economies of scale.

Mass production treats workers as little more than sophisticated machines. The fact that workers are people, and therefore are different, is seen as an obstacle to industrial production. Assembly lines require people who will do what they are told, rather than think for themselves. When most of the assembly-line workers were displaced farmers and other rural people, the industrial factories worked well. Rural people worked, as a matter of principle, no matter how they were treated. But later generations of factory workers rebelled at the human degradation of industrial work. They demanded higher wages, and worker productivity declined. Computers and robots are now replacing industrial workers, with little apparent consideration for future employment opportunities for displaced workers.

Industrialization, with its mass markets, also ignores the very real differences among preferences and needs of people as consumers. Mass marketing works best in cases where many people want, or at least are willing to accept, pretty much the same thing. The Model T Ford is a classic early example of a successfully mass-marketed industrial product. Blue jeans are another example of something that could be mass marketed to working-class people on the basis of comfort and durability. Many generic foods—beef, pork, chicken, rice, potatoes—also are examples of successful mass markets. However, such mass markets were successful because they offered lower prices than had previously prevailed in the marketplace. Consumers accepted these things not because they really preferred them, but because they were much cheaper than the alternatives.

As consumers' incomes increased, they became more able to buy the things they really wanted and less willing to settle for whatever was offered. Both cost savings and effectiveness declined as industrial principles were applied to produce goods and services for which consumer wants and needs were inherently diverse, such as in education and health care. In the case of food, packaging and advertising now claim a larger share of our food dollar than do the farmers who produce our food. There is every reason to question a system in which more money is required to convince people they want something than is required to produce the things consumers have to be convinced that they want. Why not just produce the things they really want, even if those things cost more to produce? Niche marketing is an answer to this question. Niche marketing allows people to express their uniqueness, as producers and as consumers, and thereby creates true economic value.

Uniqueness is the key to sustaining profits in farming. Uniqueness of geographically fixed resources—land, groundwater, climate, and so forth—creates opportunities to produce products that cannot be duplicated in different locations. For example, wine from grapes grown under different soil and climatic conditions are unique in quality. Successful small-scale winemakers each have a niche market made up of customers who prefer their particular wines. Niche markets for Vidalia and Walla Walla onions provide similar examples of geographic uniqueness.

Each producer within a region also has a unique set of talents and skills that cannot be duplicated by other producers, although many have not yet discovered and learned how to use them. Such uniqueness is commonly associated with those who are labeled as "craftsmen"—those who produce things of value that cannot be mass-produced. Those farmers who use unique talents to produce for unique market niches likewise are craftspeople; they produce things of value that cannot be mass-produced. There is sustainable value in the uniqueness of people.

The uniqueness of any niche market is enhanced when it is linked to personal relationships. In such cases uniqueness is not just a characteristic of people individually but also of the relationships between people. Attributes such as trust, confidence, and respect exist between people rather than within people. Positive relationships are potentially valuable assets in business as well as in personal endeavors. People like to do business with people they trust and respect and in whose commitment they have confidence. They value such relationships and are willing to reflect that value in price and other terms of trade.

Industrialization is an inherently impersonal system. Rules of trade—government grades and standards, contract law, price-reporting requirements, and labeling and advertising requirements—have been devised to minimize the inherent costs of depersonalized markets. However, over time consumers have lost confidence both in the industrial corporations who distribute their food and in the government bureaucracies that are supposed to protect them from the corporations. They don't trust either government or food corporations to ensure that the food supply is safe and wholesome or that consumers are not being "ripped off" economically somewhere in the process. Trusting relationships between farmers and customers are not only possible but are typical of successful niche marketing and are valuable assets to both farmers and their customers.

The level of profit that is sustainable is directly related to the degree of uniqueness. Substitutes exist for all economic goods and services. The degree of substitutability exists in the mind of the buyer. Some things have a lot of good substitutes and others have only a few, and even those may not be very good. The latter, those with few substitutes, are more unique. Things with a lot of good substitutes cannot command a price much higher than the price of their substitutes. Those things having few good substitutes command higher prices because there aren't many good alternatives at any price.

Others inevitably try to emulate the producers of the higher-priced profitable products. In mass markets, where things are pretty much the same, competition quickly eliminates any price premiums not associated with higher costs (unless the seller has the market power to restrict competitors). But if higher values are associated with characteristics unique to a specific producer and place of production, others simply cannot duplicate them elsewhere. The magnitude of profits will reflect the degree of uniqueness, and the sustainability of profits will reflect the persistence of that uniqueness over time.

Attempts to expand a market beyond its niche—its limited size and scope—invariably turns it into a mass market with all the associated implications. For example, if niche marketers go beyond providing accurate product information and resort to persuading customers to buy their products, they are moving into a mass-marketing strategy in which costs and competitive strategies eventually will dictate the price. Their efforts to make *more* profits will destroy the sustainability of the profits they once had. They will have responded to the common human failing of not knowing how much is enough.

The controversy over national organic standards has been as much about *whether* there should be national standards as *what* the standards should be. Organic producers for the most part have been niche marketers selling to local customers. Lacking effective state and national standards, many customers had to rely on their personal knowledge of the individual producer or local organization to ensure that products were organic. The ability to gain customer trust effectively limited an individual producer's market but also provided protection from outside competition. Effective national organic standards will eliminate both. National standards will turn many small organic niche markets into one large organic mass market, with all the associated consequences.

The industrial system is driven by the conventional economic assumption that human wants are insatiable—that we can never have enough. We can get enough of specific things but never enough of everything, which translates into never having enough money or enough profit. But sustainability is based on the assumption that we can have enough—that we will choose to leave something for others. Successful niche marketers must be willing to break out of the industrial mindset. They must be willing to decide when they have enough. They must be willing to accept the fact that each niche market is limited in size, even though the niches are not limited in number. The key to successful expansion is to find another niche rather than expand a niche beyond its natural limits.

Niche markets likewise are sustainable only if they are part of a sustainable system of production and marketing. If niche markets are pursued only for the purposes of enhancing profits, they will not be sustainable. A niche market for products produced by means that degrade the natural environment and degrade the quality of life of workers or others in the community is not sustainable. If production is not sustainable, the market is not sustainable, no matter how profitable it may appear to be in the short run. Much of the current emphasis on niche marketing is a product of the same short-run self-interest and greed that has driven mass marketing. Niche marketing becomes the key to sustainability only if it is used to enhance the economic viability of systems that are ecologically sound and socially responsible.

There are as many niche markets as there are individual wants and needs, but some niches are more promising than others. The key to finding profitable niche markets is to look where industrialization has either run its course or is being used in situations where it just doesn't fit and thus is causing problems.

In agriculture, the market niches tend to be more promising for those commodities where the marketing share of the consumer's food dollar is greatest. Pastas, baked goods, and natural cereals are just a few examples where "value adding" marketing services accounts for more than 90 cents of each dollar consumers spend. By turning such raw agricultural commodities into "value-added" consumer products, farmers have an additional 90 cents from which to glean additional profits.

Perhaps the most widespread among successful niche markets are those for fresh vegetables and fruits. Success here can be traced to very real quality differences between local, fresh produce and produce shipped in from distant regions. Harvesting, storage, and distribution methods made necessary by mass distribution systems have resulted in a significant deterioration in sensory quality of many fresh vegetables and fruits. Hard, tasteless tomatoes are a classic example. But nearly all shipped fresh produce is selected for shelf life rather than flavor. Because it is shipped unripe to reduce damage, it typically grows stale by the time it leaves the supermarket. Local producers who offer fruits and vegetables that are selected for flavor, picked ripe for taste, and sold quickly have a natural quality advantage that cannot be duplicated by industrial systems. Similar untapped opportunities likely exist for other farm commodities.

A growing concern for food health and safety represents a potential growth area for niche markets. As markets move increasingly toward globalization, consumers will have even less knowledge and less confidence in the safety of conditions under which their foods are grown and processed. They will be increasingly concerned about chemicals used in other countries that are not allowed in the United States and about sanitation standards in foreign processing facilities. Local producers who can assure local customers of high health and safety standards—including but not limited to organic production—may find increasingly profitable and sustainable niche markets.

Environmental and animal welfare concerns provide similar opportunities for niche markets. However, in these cases the customer's concern is driven more by ethics and moral values than by individual self-interest. The customer is less concerned about the healthfulness of the final product than with the methods used in producing it. This dimension of the recent organic standards debate seemed to catch the agricultural establishment by surprise. One group of potential organic customers may be concerned about food safety and health issues, but others may be more concerned about the environmental or ethical implications of conventional farming methods.

The agricultural establishment also has a hard time understanding customers who seem perfectly willing to have animals slaughtered for food but who don't want those animals to be mistreated. They don't seem to understand that it's a matter of ethics—fundamental beliefs about the right and wrong of human relationships with other animals. It's not so much a matter of animal rights as of human responsibility. People, both rich and poor, are willing to pay more to support those who share their beliefs. Producers who can find customers who share their ethical values concerning the environment or animal well-being may find sustainable niche markets.

Perhaps the most promising aspect of niche marketing is the opportunity it offers to reconnect farmers with customers through the development of personal relationships. Farmers' markets and community-supported agriculture enterprises provide prime examples. The industrial system had to separate people within families, communities, and throughout society in general in order to achieve the economies of specialized large-scale production. It separated people from each other horizontally—separating farmers within rural communities—and vertically—separating farmers from their customers. Farming systems that bring families and neighbors back together not only have economic advantages through better utilizing time and talents, but also help restore the quality of life in rural communities.

Communities require more than jobs, businesses, and social services to survive and to thrive. Communities are about people with common bonds of shared interests in the present and shared hopes and aspirations for the future. Industrialization has broken these common bonds—leaving collections of people living in the same places but with little sense of community. Local food systems can strengthen both the economic and social fabric of a community.

The most difficult challenge in sustainable niche marketing is to break out of the old paradigms of industrial mass marketing. Many who start out as niche marketers eventually fall back into old mass-marketing ways of thinking about pricing, advertising, cost cutting, convenience, product appearance, shelf life, and so on. Niche marketing requires new, unconventional ways of thinking and acting.

Robert Kriegel in his book *If It Ain't Broke . . . Break It*, outlines some "unconventional wisdom" that may prove valuable to would-be sustainable niche marketers.[3]

Believe in providence! Ride the waves of change! Niche marketing is a wave of the future. Even the large corporations are moving away from mass marketing and trying to tailor production to more narrow market niches. But the advantage of being big comes from being able to do a lot of the same stuff, over and over. Small producers can compete in niche markets. Small may well be the wave of the future. Be willing to ride the wave.

Don't compete! Change the game instead! If you want to compete with the big boys, you better be big and you better be mean. Corporations have access to unlimited capital, and they have no heart or soul. If you don't have a lot to invest and you aren't willing to sell your soul, you better stay out of the big boys' game. Change the game to one that the big boys can't play. Succeed by being small and unique rather than big and tough.

Trust the unexpected! Plan to change your plan! Sustainable agriculture and niche marketing are new arenas for most farmers. What we see today in sustainable farming and niche marketing is analogous to the old steam threshers that lumbered down the road in the early days of industrial agriculture. There is a lot yet to be learned, a lot of changes to be made, and a lot of things will happen that are totally unexpected. Be prepared to roll with the punches—plan on changing your plans.

Don't be realistic! Dreams are goals with wings! Many if not most people in the agricultural establishment believe sustainable agriculture is an idealistic dream. Many say that niche markets may work for a few folks in special situations but is not something to be taken too seriously. But sustainable agriculture—rather than biotechnology, precision farming, franchise farming—may well be *the* farming of the future. Niche markets—rather than mass markets—may be *the* markets of the future. Why should we be realistic when no one knows for sure what is really going to happen? Why not put wings on your goals and dream?

Light a fire in your heart! Passion is contagious! If you are going to try something really different, you had better do it with passion. New things take a lot of creative thinking and a lot of hard work. Other people are going to disagree with you and laugh at you. If you don't start with fire in your heart, you had better not start. Others will dump water on your flame, so you are going to need friends who will relight your fire. Passion for life and work is the best way to bring others to share your cause. Light a fire in your heart and fuel it with passion, and you just might make it work.

Joy pays off! Learn to play . . . to win! Life is not a destination; it is a process. The destination for all is death—not life. Most of us spend a good part of our life at work. If there is no joy in your work, there is no joy in a big part of life. If there is little joy in your life, there is little chance for success—regardless of how much money you earn being miserable. If you really want to succeed, you must put joy in your work and your life. Find ways to turn work into play. Then you can play to win without losing.

Sustainable agriculture is a new way of thinking about farming, about work, and about life. In a sense, sustainable agriculture is about new life—and life begins *outside the eggshell.* Marketing in the niches is about marketing in the cracks of the old mass-production, mass-marketing food system. Even if it ain't completely broken yet, we can break it! Marketing in the niches, not mass marketing, is the key to sustainable farming.

⌐⌐ Presented at the Annual Fall Conference of Resilience, Inc., Creola, Ohio, October 10, 1998.

15

Local Organic Farms Save Farmland and Communities

One thing we all have in common is our dependence on the land and on each other. We are still as dependent on the land for our daily sustenance and survival as when all people were hunters and gatherers. Our dependence is less direct and our connections more complex, but human life, like all life, is still critically connected to life in the soil and to the farmers who help nurture life from the soil. Yet despite the importance of land to us individually and collectively—to our communities, our nations, and the world—the health and productivity of farmland almost everywhere is in peril. Farmland is at risk from soil erosion, agrochemical contamination, salinization, overapplication of animal wastes, and a host of other threats posed by the industrialization of agriculture. But it is also imperiled by poorly planned development, especially in urbanizing areas. On urban fringes everywhere, complex sets of economic and political forces are driving the conversion of some of the best farmland in the world into urban residential and commercial developments.

In the United States alone, between 1992 and 1997, more than 11 million acres of rural land were converted to residential and commercial use—and more than half of the land converted was agricultural land. On average, more than 1 million acres of U.S. farmland were developed each year during this period.[1] This rate was more than 50 percent higher than the rate reported in the previous decade, and there is reason to believe that the rate of conversion has accelerated since 1997. While a million acres per year may seem small in relation to the 930 million acres of U.S. farmland, an acre claimed by urbanization is an acre irretrievably lost from human food production.

In spite of our proclamations of personal independence, people are as dependent on each other as they are on the land. We depend on each other for life and for quality of life, just as much as when people lived in clans, tribes, or agrarian communities largely isolated from the rest of humanity. Our relationships have become less personal and our dependencies more complex and less clear, but few of us could prosper or even survive for very long without other people. Perhaps even more important, millions of us suffer a diminished quality of life because we have lost our sense of personal connectedness to other people. Yet people everywhere continue to leave caring communities and accept urban isolation in their quest for greater individual wealth and general economic prosperity. And we question why society seems less trusting, less civil, more abrasive, and more violent.

The current threats to both land and to communities ultimately stem from the failure of the economy, particularly the agricultural economy, to evolve in ways necessary to meet the diverse needs of people in an increasingly crowded world. For example, the currently dominant industrial approach to agricultural production was driven by a perceived need to produce more food and fiber for a growing world population with fewer farmers and at lower costs. We had to free people and resources from food and fiber production so they would be available to create modern industrial economies.

Today, farmers are told by the agricultural establishment that the world's consumers demand increasing quantities of low-priced foods of consistent quality, and that a corporately controlled, vertically integrated industrial agriculture is the only means by which this demand can be met. The new high-tech, biotech, vertically integrated global food chains require large-scale, specialized, standardized systems of agricultural production. Its advocates believe this industrial approach to farming is dictated by the impersonal forces of free-market economics, and thus is inevitable. The discipline of industrialization is enforced through the farmer's economic bottom line. If farmers expect to survive, they are told, they have no other choice.

However, residents of rural communities are voicing increasing concerns about current trends in agriculture, at least in U.S. agriculture. They are concerned about the pollution of groundwater and streams with pesticides and fertilizers from specialized cropping systems and with manure runoff and spills from large-scale confinement animal feeding operations. Many people now know that agriculture has become the number one nonpoint source of stream pollution in the United States, and a growing "dead zone" in the Gulf of Mexico is attributed largely to agricultural pollution sources.[2] The agricultural establishment typically denies blame for these problems, but no one can reasonably deny the reality of growing public concerns regarding the ecological impacts of agricultural industrialization.

Rural residents are concerned also because of the negative impacts industrialization has had on rural communities all across the United States. As farms became larger, farm families became fewer, and it takes people, not just production, to support a community. Larger farms also have tended to bypass their local communities in buying inputs and marketing products. Today, the management, if not ownership, of even the largest farming operations is being consolidated under the control of giant multinational agribusiness corporations through comprehensive contractual arrangements. The few contract producers remaining hold little resemblance to traditional independent family farmers. The agricultural establishment may argue that the abandonment of rural places is inevitable, but no one can reasonably deny that the demise of small and mid-sized family farms is destroying rural communities.

While still a minority, a growing number of consumers also are becoming concerned about the wholesomeness, nutrition, and safety of foods produced by industrial agriculture. Organic food has become the fastest-growing segment of the retail food market. Consumers don't want chemicals in their foods. Parents have demanded that soft drink machines and fast foods be removed from their schools. They want nutrition in their children's diets. A few customers have sued fast food restaurants for causing their obesity, accusing the corporations of peddling addictions. More consumers are buying more of their food locally, so they won't have to overcook their hamburgers and eggs to avoid salmonella poisoning or worry about *E. coli* contamination or "mad cow disease." The agricultural establishment continues to praise the safety, nutrition, variety, and value provided by the modern food system. However, many consumers have lost confidence in the integrity of the corporate industrial system and in the ability of the government to regulate it.

Organic farming emerged from obscurity to become a popular food trend as a consequence of these growing concerns about the impacts of industrial agriculture on human health and the natural environment. People were concerned about the health effects of not just agrochemical residues in their foods but also agricultural pollution of groundwater and streams. Standards for organic production prohibited the use of the agrochemicals linked with environmental pollution and prohibited the use of pesticides linked with health risks in foods. Many consumers were willing to pay substantial price premiums for organically produced foods, and farmers were willing to produce them. And at a rate of growth of more that 20 percent per year during the 1990s, the market for organic foods has been doubling every three to four years.

Organic farmers historically not only have taken better care of the soil, they also have tended to be smaller independent farmers who are more supportive of rural communities. A 1998 survey of the Organic Farming Research Foundation indicated that nearly 90 percent of U.S. organic farms were single-family operations or family partnerships.[3] More than 60 percent were full-time farming operations, even though the average size of an organic farm was only about 140 acres—just over one-third as large as the average U.S. farm. Thus, organic farming in the United States was still dominated by small family farms, at least in terms of farm numbers. There was reason to hope that organic farming might become mainstream farming, and thus, organic farming might restore fertility to the soil and organic farming families might restore life and health to rural communities. Organic farming might save both farmland and communities.

Today, however, even organic farming is becoming industrialized. Potential profits from the rapid growth in markets for organic foods eventually attracted the attention of the large food corporations. Large food processors and retailers found it difficult to deal with the diversity of organic standards and certification programs that existed among different groups of farmers in different regions, both within and among nations. They encouraged organic farmers to adopt uniform standards for national organic certification and to harmonize standards among nations. Organic farmers were led to believe uniform standards would give them greater access to new mass markets. On the surface, standardization seemed like a good idea.

Instead, uniform national and international standards simply facilitated the specialization, standardization, and consolidation of the organic food production and marketing. With uniform standards, large-scale specialized producers who could meet minimum standards and lower costs could now produce and ship large quantities of organic products across nations and around the world. This industrialization of organics left smaller independent organic farmers, like smaller independent conventional farmers, struggling for economic survival.

Most economists and many agriculturalists seem unconcerned about either the industrialization of agriculture or the industrialization of organics. To economists, it is just a matter of farmers and corporations minimizing their costs of production. Consumers benefit from lower-cost food, and the farmers who are displaced ultimately find employment elsewhere. Many economists seem to believe that we still have too many farmers to allow efficient food production. They claim that smaller independent farmers have been able to survive only because of protectionist farm subsidies and that removing these subsidies will result in a more efficient allocation of land, labor, and capital. Higher profits for corporate investors and economic benefits for contract producers will more than offset losses in farm income, they say, leaving the agricultural economy stronger than before. To the agricultural establishment, the organic market is nothing more than a market niche made up of emotional, misinformed, affluent consumers who fail to appreciate the technological miracles of the modern food system.

Unlike most economists, I am very concerned about what's happening to the food system in general and to the organic market. Being an economist is no excuse for ignoring ecological and social reality. How can agriculture meet the food and fiber needs of a growing world population if we destroy the natural productivity and regenerative capacity of the land? Economists generally assume that we will find substitutes for anything we use up and will fix any ecological or social problems we create, but these are simply beliefs with no logical, scientific support in fact.

What is the net benefit of an agriculture that meets the physical needs of people but separates families, destroys communities, and diminishes the overall quality of life within society? How can it possibly be right or good to defile the earth, even if it is profitable to do so? Economists simply don't consider the social, psychological, or ethical consequences of the things people do to make money. Economics treats such things as social or ecological externalities, which may impose irrational limits or constraints on the legitimate pursuit of wealth.

Obviously, I am not a typical economist, although I have been an economist for more than thirty-five years, serving on the faculties of four major U.S. universities. For nearly half of those years, I thought pretty much like most other neoclassical economists. I believed the market was always right, I believed that bigger was generally better, and I believed in the conventional wisdom of farming for the economic bottom line. However, my beliefs were shaken by the farm financial crisis of the 1980s, when many of the farmers who lost their farms had been doing the things we economists had recommended. After a period of serious questioning, I eventually concluded that we economists, and many others in the agricultural establishment, were simply out of touch with reality. We had been trying to transform farming into something that it was not and could not be. We had treated the farm as if it were a factory without a roof and fields and feedlots as if they were biological assembly lines. We had encouraged farmers to specialize, standardize, and consolidate, as if farming were a manufacturing process, simply transforming inputs into outputs. We had been treating farms, which are complex biological and social organisms, as if they were nothing more than sophisticated machines.

Luckily, at about this same time, sustainable agriculture was making its way onto the agricultural scene in the United States. The more I learned about sustainable agriculture, the more I realized that it might be the answer to my growing questions about why the kind of agriculture I had been promoting wasn't working. But more importantly, in sustainable agriculture I was beginning to see an agriculture that could actually be good for farmers and rural residents, as well as consumers.

Through my interest in sustainable agriculture, I acquired my first real knowledge of organic farming. I learned from talking with committed organic farmers that organic farming meant more than just farming without commercial chemicals. It was a philosophy of living, as well as farming, in partnership and harmony with nature. But I didn't see most organic farmers promoting the organic philosophy, either with their customers or in their professional relationships. The emphasis seemed to be almost completely on farming without synthetic pesticides and fertilizers. I knew from history that farming without chemicals provided no assurance of sustainability. Great civilizations had fallen because their agricultural systems had not been sustainable, and their farmers obviously had farmed without agricultural chemicals. The fact that commercial organic production was more environmentally sound than conventional industrial farming didn't necessarily make it sustainable. I considered organic farming—along with biodynamic farming, holistic resource management, and permaculture—to be a legitimate approach to sustainable agriculture. However, I wasn't sure that any of these approaches were actually sustainable, at least as they were currently practiced.

Apparently, my early perceptions of commercial organic farming were shared by those who eventually shaped national and international organic standards. The official certification documents articulate commendable goals and principles for organic production, but the actual standards deal almost exclusively with specific allowable and nonallowable inputs and practices and specific requirements that must be met to achieve certification. The new organic standards barely hint at the philosophy of organic farming I had learned from the organic pioneers of the 1960s and 1970s. It was relatively easy to see in advance that national and international certification would lead eventually to the industrialization of organic farming.[4]

The optimism of the 1990s quickly turned into despair for many organic farmers. As large-scale corporate organic producers linked up with large corporate wholesalers and retailers, smaller independent organic farmers lost access to the growing market they had spent years developing. A few of the more successful organic operations were bought out by the new corporate players, but most lost markets and many lost their businesses. Organic premiums at the farm level shrank, as larger producers who could meet minimum government standards for lower costs gained an increasing share of the organic market.

Previous relationships of trust among organic farmers, buyers, and consumers were replaced by government rules and regulations. The organic farmer's philosophical commitment to stewardship and integrity was replaced by the government's assurance of conformity. How could independent family farmers ever expect to compete with the economically and politically powerful food corporations? The situation seemed hopeless; there seemed just cause for feelings of despair. The earlier hopes that organic farming might restore health to the land and renew life in rural communities grew dim.

Organic farmers now faced a dilemma similar to that of conventional farmers—should they try to get larger, give in to corporate control, or give up and get out of farming? The solution to this dilemma is not found by responding to feelings of hopelessness with acts of recklessness, but instead, by acting with the calm assurance that it is always darkest before the dawn. Independent organic farmers are not going to supply the Dominions, A&Ps, or Wal-Marts of the world, and probably not even the large specialty organic supermarkets. Organic price premiums are probably not going to return to levels of the 1990s. But markets for locally grown organic foods are growing faster than ever. And smaller independent organic producers have a distinct advantage over the larger producers in serving local markets. It's time to move beyond organics to organics-plus, to begin building a new network of locally oriented, community-based organic food systems.

An increasing number of consumers in the industrialized nations of the world want something more than organic; they also want to know where their food is grown and who grew it. They are concerned about freshness, flavor, nutrition, and overall food safety and quality, not just pesticide contamination. They are concerned about the impacts of their food choices, not only on the natural environment but also on the health and well-being of farmers and farm workers.

These discriminating consumers want food produced by local farmers, by real people with integrity whom they can trust. And if they can't buy locally, they still want food produced by farmers who are known and trusted by customers in their respective communities, by farmers who must rely on integrity as much as productivity for their advantage. Marketing opportunities exist for local organic farms all across America, in rural, urban, suburban, and urbanizing areas. Regardless of the area, farmers—real people—have a distinct market advantage over corporations in local organic food markets.

The emergence of a new food culture in the United States is validated by the Hartman Report, a nationally respected survey of U.S. households that explores the linkage between food purchases and environmental attitudes.[5] The report identified two groups of consumers, the "true naturals" and the "new green mainstream," which together make up about 28 percent of the U.S. population, as prime markets for organic and sustainably produced foods. This new food culture appears to be just one part of a broader, more inclusive new American culture. A growing body of evidence indicates that somewhere between one-fourth and one-third of Americans want foods that are fundamentally different from those available in American supermarkets, franchised restaurants, and commercial eating establishments today.[6] The people of the new food culture care about food quality and safety, but they also care about the social and ecological consequences of their food choices.

Organic farmers can play an important role in helping to create this new food culture by simply returning to the roots of organic farming. The roots of organic farming are in philosophy, in questions of good and bad or right and wrong, which simply cannot be encoded in lists of inputs or descriptions of farming practices. Organic farming is fundamentally about the rightness of relationships among people and between people and the soil. Eliot Coleman, a writer and organic farmer, calls true organic farming "deep-organic farming" in the same sense that "deep ecology" asks not only how people *do* relate but also how people *should* relate to the other living and nonliving things of nature.

Coleman writes, "Deep-organic farmers, after rejecting agricultural chemicals . . . try to mimic the patterns of the natural world's soil-plant economy. . . . Shallow-organic farmers, on the other hand, after rejecting agricultural chemicals, look for quick-fix inputs. Trapped in a belief that the natural world is inadequate, they end up mimicking the patterns of chemical agriculture."[7] Deep-organic farming, like deep ecology, is based on the understanding that we humans are not separate or isolated but are integrally interconnected with each other and with the world around us. Health of the soil, health of people, and health of society are integral aspects of the same whole. We are all part of the same flow of energy, the same web of life.

The father of biodynamic farming, Rudolph Steiner, in a landmark series of lectures in 1924 wrote, "A farm is healthy only as much as it becomes an organism in itself—an individualized, diverse ecosystem guided by the farmer, standing in living interaction with the larger ecological, social, economic, and spiritual realities of which it is part."[8] In this sense, organic describes the organization of the farm as a living system, as an organism. Steiner considered the rightness of relationships among the farm, farmer, food, and eater to be divinely determined. He was concerned that food grown on the increasingly impoverished soil of conventional farms could not provide the inner sustenance needed for spiritual health.

Early advocates of organic farming believed that human health was directly connected to the health of the soil. Soil scientist William Albrecht wrote in 1952, "Human nutrition as a struggle for complete proteins goes back . . . to fertile soils alone, on which plants can create proteins in all completeness."[9] Organic pioneer and publisher J. I. Rodale wrote, "The *organiculturist* farmer must realize that in him is placed a sacred trust, the task of producing food that will impart health to the people who consume it. As a patriotic duty, he assumes an obligation to preserve the fertility of the soil, a precious heritage that he must pass on, undefiled and even enriched, to subsequent generations."[10]

Sir Albert Howard and other organic pioneers also emphasized permanence as the core principle of organic agriculture. Howard began his book *An Agricultural Testament* with the assertion, "The maintenance of the fertility of the soil is the first condition of any permanent system of agriculture."[11] In his opening chapter, he contrasted the permanent agriculture of the Orient with the agricultural decline that led to the fall of Rome. He wrote, "The peasants of China, who pay great attention to the return of all wastes to the land, come nearest to the ideal set by Nature. They have maintained a large population on the land without any falling off in fertility. The agriculture of ancient Rome failed because it was unable to maintain the soil in a fertile condition." Of the fall of Rome, historian G. T. Wrench wrote, "Money, profit, the accumulation of capital and luxury, became the objects of landowning and not the great virtues of the soil and the farmers of few acres."[12] Howard concluded, "The farmers of the West are repeating the mistakes made by Imperial Rome."

The historic purpose of organic farming was permanence—to ensure the sustainability of agriculture, and through agriculture, the sustainability of human society. Only living organisms, including living organizations, have the capacity for regeneration, and thus, the capacity for permanence. Nonliving systems inevitably tend toward entropy, "the ultimate state reached in degradation of matter and energy; a state of inert uniformity of component elements; absence of form, pattern, hierarchy, or differentiation."[13] Living systems, however, are capable of capturing and storing solar energy to offset this inevitable degradation of matter and energy. Thus, permanence is inherently dependent on healthy, living organic systems of production. Only organic systems are capable of restoring the energy and matter that is degraded in the natural tendency toward entropy. Only living systems are capable of permanence.

Permanence requires sustainability—an ability to meet the needs of the present without compromising the future. A system of farming that destroys the natural productivity of the soil cannot sustain its productivity. A system of farming that fails to meet the needs of all, both producers and consumers, equitably, will not be sustained by that society. And a system of farming that cannot be made financially viable will not be pursued, no matter how ecologically sound or socially just it may be. The requirements for sustainability can be met only by regenerative living systems. The principles that guide living systems cannot be captured in a set of written standards, and thus cannot be imprisoned in a rigid set of rules and regulations. These principles must be written in the hearts, minds, and souls of people, not just as farmers but also as consumers, as citizens, as morally responsible human beings.

Deep-organic farming ultimately depends on people making a personal commitment to maintaining the health and productivity of self-renewing, regenerative, living ecosystems, societies, and economies. Such personal commitments require a sense of personal connectedness to people and to place. In the words of Wendell Berry, "Farming by the measure of nature, which is to say the nature of the particular place, means that farmers must tend farms that they know and love, farms small enough to know and love, using tools and methods that they know and love, in the company of neighbors they know and love."[14] Deep-organic farming depends on personal relationships of integrity and trust among farmers, farm workers, eaters, and citizens within local communities.

This commitment to personal connectedness, to relationships of integrity and trust, is precisely what those in the new food culture are searching for and are willing to commit their time and dollars to finding and nurturing. Their commitment to supporting local foods and local farmers is a commitment to establishing personal connectedness to people and place. The most common examples of places where this connectedness occurs are farmers' markets, roadside stands, csas—where farmers and their customers meet face to face. However, integrity

and trust can be built anytime there is a sense of personal connectedness between those who produce food and those who eat
it. Such connections are far easier to establish and maintain in
situations where farmers, processors, retailers, and customers
all live in geographic proximity—in local markets.

Community-based food systems are not intended to make
local communities self-sufficient in food production, any more
than sustainable agriculture is intended to make farms self-sufficient. The goal is not independence but interdependence—dependence by choice, not necessity. The purpose in both cases is
to find ways to work and live in harmony with nature, including
human nature, to build positive relationships among people and
between people and the earth. A local, community-based food
system encourages and supports production of foods uniquely
suited to specific ecological and cultural niches, as a means of
achieving this harmony. It also encourages and supports local
consumption of local foods, in the belief that eating foods produced in the places where we live, by people we know, is an act
of integrity and value. It encourages people to come together,
to create a sense of community, around food.

Local food systems of integrity can be connected with other
local food systems of integrity through intercommunity relationships of integrity, forming national and global food networks of integrity. The fundamental purpose in creating such a
food network is to reconnect us to the earth and to each other.
However, connectedness does not require that consumers eat
only food produced locally or that farmers should sell all of
their products locally. Connectedness simply requires a sufficient commitment to the local community to reconnect us
with each other and with the land and to enhance our quality
of life.

No other type of farming is more clearly capable of helping to create this new community-based food system than is
deep-organic farming. In fact, deep-organic farming requires
personal commitment to relationships of integrity among peo-

ple and between people and the land. Such relationships must characterize sustainable community-based food systems of the future.

Conventional farming lacks both the ecological and social integrity that the new food culture demands. Industrial organic farming makes no social or ethical commitments to farmers, families, communities, or to the land. Corporations are incapable of personal relationships because they are not people. Only real people who are committed to relationships of rightness are capable of maintaining relationships of integrity and trust.

Perhaps even more importantly, nowhere is the potential for local organic farming greater than in urbanizing areas. Deep-organic farming does not threaten the health and productivity of the land or the health and environment of one's neighbors. Instead, deep-organic farming depends on the natural productivity of the land and the support of the community for its productivity and profitability. Deep-organic agriculture is a land-friendly, people-friendly approach to farming. It provides a means of building positive relationships of respect and trust between farmers and their new neighbors in urbanizing areas. With this approach to organic agriculture, farms and residential developments can coexist, and even build new thriving rural communities, by sharing the same spaces for farming and development.

Organic farms can be good places to live and raise a family and good places to have in the neighborhood. Productive farmland might well take the places of green belts and golf courses in urban residential developments, with residences clustered in the more aesthetically pleasing and less productive ridges, hillsides, and wooded fringes. Individual residential lots might include ownership of a share of the farmland with the community in common. The former farmer-landowner could realize the full economic value from the land for development while retaining the privilege of farming the land on behalf of his or her new community members. Residents could be afforded many of the amenities of life on a family farm without having to learn to do the farming. Permanent land-use restrictions could be placed on such developments to prevent future reversion of the farmland to more-intensive residential development.

In other situations, high-density residences could be clustered in less-desirable farming areas, creating small but efficient urban areas surrounded by open farmland. Developers of the urban centers could be required to buy development rights from the surrounding areas, leaving productive farmland to be farmed while creating desirable places for people to live.

Potential positive solutions to farming and living in an increasingly crowded world are endless. The key is the pursuit of harmony through sustainability in farming and living, which requires ecological integrity and social responsibility to ensure economic viability. Mutual respect and consideration arises from the realization that caring for neighbors and caring for the earth, as we care for ourselves, is simply a more desirable way to work and to live.

These concepts are not radical or new. They have been used in Europe for centuries, where farmland surrounds small villages in which even the farmers live. These so-called cluster housing developments also have been popular in some rural areas of America, particularly in the eastern United States. Surveys have shown that residents generally rate them very highly as places to live, and they have maintained their property values well.[15] The emergence of a new, locally oriented deep-organic food system could not only save farmland from industrial degradation and urbanization but also could save existing rural communities and help build healthy new communities in urbanizing areas. Local organic farms can save farmland and communities.

This new vision may at first seem idealistic—little more than an unattainable utopia. But we should remember that the current industrial food system was not built all at once by some fiat or decree, but was built by individuals, one person at a time.

One by one, as consumers changed what they ate and where they bought it, one by one, as farmers changed what they produced and where they sold it, and one by one, as processors and distributors changed how and where they operated, the food system was transformed from organic and local to industrial and global. So, one by one, as farmers, consumers, processors, and distributors make different choices, the food system can just as easily be transformed from industrial to organic and from global to local. Today, those choices are being made.

The noted anthropologist Margaret Mead said, "Never doubt that a small group of thoughtful, committed citizens can change the world. Indeed, it's the only thing that ever has." People of the world will continue to be dependent on the soil, and on the people who nurture life from the soil, for as long as there are people in the world. And people of the world will continue to be dependent on caring communities, where people truly care about other people, for as long as there are people in the world. One by one, people are beginning to recognize these dependencies and are beginning to reflect these realities in their choices. Never doubt that deep-organic farmers, who care about the land and care about people, can save farmland as well as rural communities. Indeed, they are the only ones who can.

¦ ¦ Presented at "Local Organic: A Global Solution," 2005 Organic Conference at the University of Guelph, Ontario, Canada, January 20–23, 2005.

16

The Triple Bottom Line of Farming in the Future

American agriculture is in crisis. Contrary to common belief, a crisis is not necessarily a bad situation. The dictionary defines crisis as "a crucial time or state of affairs whose outcome will make a decisive difference, for either better or worse."[1] The crisis in American agriculture today is reflected in the disappearing middle class of farmers. Within this crisis are great risks but also great opportunities. Whether the ultimate outcome of this crisis is for better or worse is yet to be determined.

As the national research and education project "Agriculture of the Middle" describes it, American farms have followed two new paths over the past several decades.[2] On one path, giant corporate agribusinesses have established contractual arrangements with large, specialized producers to produce bulk commodities for both domestic and global markets. On the second path, small-scale diversified farms have thrived by successfully adapting to marketing food products directly to customers. The traditional full-time family farm has been left with little choice other than to choose one of these two paths, or the path out of farming.

According to the 2002 Census of Agriculture, farms with sales of over $500,000 per year, by most accounts large farms, accounted for just 3 percent of total farm numbers but 63 percent of the total value of agricultural production.[3] Farms selling less than $100,000 per year, which most would agree are small farms, accounted for 85 percent of all farms but only 10 percent of the total value of agricultural production. This left the farms in the middle, those with sales between $100,000 and $500,000—not too large, not too small—with only 20 percent of the farms and 27 percent of the total value of production. The 2002 census simply confirmed the trend of the past several decades. The middle class of farmers is quickly disappearing. Most of the farming middle class produce neither bulk commodities under corporate contracts nor food products for niche markets. By 1997, 63 percent of the large commodity producers already were specializing in single commodities and were producing under corporate contracts.[4] The percentage today is likely much higher. So far, very few of the farming middle class have been willing to accept producing for direct markets or niche markets as *real* farming. They seemingly would rather give in to corporate control or get out of farming than to admit to having to take the second path.

The consequences of the disappearing middle are critical, not just to the future of farming in America but also to the future of America.[5] The trend toward large-scale operations, operating under comprehensive corporate contracts, is promoted by those in the agricultural establishment as the only means by which American producers can compete in the global marketplace. However, the continued industrialization of agriculture is provoking increasing environmental and social concerns among rural residents. The agricultural establishment has attempted to label local opponents of "factory farming" as emotional, misinformed radicals who naively oppose all technological and economic progress. But the growing numbers of opponents and the growing body of scientific evidence documenting the negative environmental and social impacts of factory farms give increasing legitimacy to their concerns.[6]

More recently, however, concerns have focused on the economic future of industrial agriculture, specifically on whether American commodity producers can actually compete in a global economy. According to a 2004 USDA report, "A decade ago, a scenario in which the value of U.S. agricultural imports would someday exceed that of U.S. exports seemed farfetched. Today, the improbable has become probable. Between 1996 and

2003, the agricultural trade surplus shrunk from $27.3 billion to $10.5 billion. Although U.S. agricultural exports continued to rise, imports were increasing nearly twice as fast."[7] These rising imports were not for coffee, bananas, or exotic foods that can't be produced in the United States. Instead, Americans increasingly are eating imported basic foods, particularly vegetables and fruits. The USDA analysts went on to concede that if current trends continued, the current trade surplus would become trade deficit by the end of the decade.

The inability of American farmers to compete in world commodity markets should come as no surprise. Costs of farmland and farm labor in other major exporting countries, such as Argentina, Brazil, and China, are a fraction of land and labor costs in the United States. And increasingly, multinational agribusinesses are investing their capital and applying their technologies in those countries where total production costs are lower than in the United States. Increasing environmental restrictions on large-scale industrial agriculture in the United States is often cited as an additional reason for agribusiness corporations to locate their production operations elsewhere. And rural residents in America are becoming more rather than less concerned about protecting their environment.

University of California economist Steven Blank outlined the economic logic of this loss of competitiveness in his recent book, *The End of Agriculture in the American Portfolio.*[8] The World Trade Organization's agreement in July 2004 to liberalize world trade in agricultural products adds further credibility to his prediction that America eventually will be forced to abandon production of agricultural commodities in favor of higher-valued uses of land and labor.[9] Realistically, America is not going to completely abandon agriculture, but at some time in the future we could easily become as dependent on the rest of the world for our food as we are today for our oil. For many of us, this is cause for serious concern.

Most economists, including Blank, see nothing to be concerned about in the disappearing middle class of farmers or the end of American agriculture. To economists it is a just a matter of allowing the free markets to work. They claim Americans will actually be better off if they can have more food at a lower cost by importing it rather than growing it domestically. Economic gains for consumers and corporate investors will likely more than offset losses to farmers and rural communities, so they aren't concerned.

However, I think there is something very wrong in American agriculture today. The disappearing middle class and the ultimate end of American agriculture may make economic sense, but it just doesn't make common sense. I simply don't believe that a bigger economy necessarily results in a better society or that a more economically efficient agriculture necessarily results in a better food system. I don't believe the demise of family farms, the degradation of the rural environment, and the decay of rural communities can be so easily justified as simply declaring them the inevitable consequences of a free-market economy.

But why would anyone pay any attention to me? The agricultural establishment—the recognized experts on agriculture and economics—tell a completely different story. If farmers adopt the new technologies, if they are willing to take their place in the corporate global food chain, and if they continue to work hard and work smart, they are told they will surely succeed. Those of us who are concerned about the corporatization of American agriculture are simply out of touch with reality, they say. We are labeled as twenty-first-century Luddites, opposed to all technological progress, or at least naively nostalgic for a return to farming of the past. But why should anyone pay any attention to me?

First, I am an economist and have been one for more than thirty-five years. I have been a professor of agricultural economics at the major state agricultural universities in North Carolina, Oklahoma, Georgia, and Missouri. I grew up as a farmer—on a small dairy farm in southwest Missouri. I was a member of the Future Farmers of America, and I believed in the future of farming. I also have been a businessman. My younger brother and I operated a small restaurant during my last year in high school and I worked in management for three years for a major meatpacking company—Wilson Foods—after graduating from college. Perhaps most importantly, I spent three-fourths of my life and half of my professional career believing and teaching the very things that the agricultural establishment is extolling today. I know where these folks are coming from because I have been there.

I used to tell farmers they were going to have to become sharp financial managers, smart personnel managers, and astute marketers, because the only farmers with a future were those who saw farming as a business rather than as a way of life. I cautioned farmers to separate farm business from family business and not allow family matters to be an economic drag of the farm. I believed the family farm was a thing of the past, not the future. I was an unabashed advocate of farming for the economic bottom line—period.

However, during the farm financial crisis of the 1980s, I began to sense that something was terribly wrong. I began to question whether there really was a future in farming. Many farmers had borrowed heavily at record high interest rates to expand production to meet booming export demand during the 1970s, only to see exports dry up, commodity prices plummet, and record farm profits turn into disastrous farm losses. The agricultural establishment at the time chastised these farmers as poor managers who should have known better than to borrow so much, or at least should have known how to survive the inevitable hard times of farming. However, I discovered that the farmers who were in the biggest financial difficulty had been doing the things that the agricultural establishment, including myself and my colleagues, had been telling them they should do. It would have been easier to deny it, and many did, but I came to realize that I had been much more a part of the farm problem than a part of the solution.

I had an opportunity, during those hard times, to sit across the table from several farm families in trouble and try to help them find some way out—short of suicide. I was working in Georgia at the time, a state where many farmers who wanted to borrow a little money had been encouraged by government loan officers to borrow a lot of money—the conventional wisdom being that small farms couldn't survive. In talking with these farmers, these real people, I began to understand that a family farm is much more than a business. The true family farm is a part of the family and the family is a part of the farm; the two are inseparable. Losing a family farm is like losing a member of the family or losing one's self; perhaps that's why so many farmers' thoughts turned to suicide at the prospect of losing their farm.

Equally importantly, I learned that most true family farmers were not in severe financial difficulty, even though all were feeling a financial squeeze at that time. Many family farmers had not followed the advice of us so-called experts. They were not overly specialized; they had maintained some diversity of enterprises, and some enterprises were still profitable. They had minimized their dependence on costly chemical inputs and farm equipment, so their cost-price squeeze wasn't quite so tight. They had not bought land to expand their operations, so their debts were more manageable. The farmers we economists had branded as laggards—resisters of new technologies and new ideas—were at least coping with one of the most severe economic farm crises of the century.

Over time I began to understand that a farm is not a factory, plant and animal production is not a mechanical process, and thus, real farming is fundamentally different from working on an assembly line or managing a factory. Farming isn't just about minimizing costs or maximizing profits; it's about nurturing and caring for living things—plants, animals, people, and even the wild things of the fields and forests and living things in the soil. A farm is not just a bottom-line business; it is far more. The family nurtures the farm and the farm nurtures the family, and the family nurtures, and is nurtured by, the biological and social communities of which it is a part.

At about the same time, the new low-input sustainable agriculture (USDA) program was being initiated by USDA. I realized that dramatic change was needed in American agriculture and thus I needed to make some dramatic changes in my professional life. I was open to any opportunities that sustainable agriculture might offer and eventually was able to secure a contract to work for the USDA LISA program. Thankfully, in sustainable agriculture I eventually found a reason to again believe in the future of farming.

I decided to return to Missouri, my home state, to carry out my new sustainable agriculture project. My first real understanding of sustainable agriculture was that of a balanced approach to farming. Missouri had a highly successful extension program back in the 1950s that focused on balancing farm profitability, soil conservation, and family living; it had been called the Balanced Farming program. The program had been driven by the need to increase farm income but without degrading the land or the quality of family life. Sustainable agriculture, on the other hand, was being driven more by the environmental concerns being raised by the industrialization of agriculture. But the needs for farm income and for a desirable quality of farm and rural life were still there.

People were beginning to understand that an agriculture that degraded the land and polluted the natural environment simply could not sustain its productivity over time. People were also beginning to understand that an agriculture that couldn't meet the needs of society—not just as consumers, but as farmers, rural residents, and people in general—would not be supported by society, and thus was not sustainable. Everyone still understood that agriculture had to be profitable, at least periodically, if farmers were to survive financially. So farming sustainably was about finding balance and harmony among the ecological, economic, and social aspects of farming. Certainly, it was about meeting the needs of the present while leaving opportunities for the future, but to me it was more just a common-sense way to farm.

Through my work with a new breed of farmers, I discovered the second path to the future of agriculture. Admittedly, most of these new farmers have smaller farming operations than do their conventional industrial counterparts. Their farms also tend to be more diverse, often integrating crop and livestock production. Many of these farmers market directly to local customers whom they know personally. But these new farmers are set apart from commodity producers, not so much by their size, products, or markets, as by their basic approach to farming and their philosophy of life. The size, products, and markets are simply a reflection of their values and their philosophy. They are farming not just for profits but for a better overall quality of life—and many are finding it.

I now have new hope for the future of farming in America because of what I see in these new American farmers. Since retiring in February 2000, I have had the privilege of speaking about issues related to sustainable agriculture at thirty-five to forty different venues each year. Most of these are conferences attended mostly by farmers who are interested in sustainable agriculture. Both the number of such conferences and conference attendance seem to grow each year. I never pass up an opportunity to visit with farmers wherever I go, and most of what I know about sustainable farming today I have learned from these new farmers.

The second-path farmers I have met along the way are very different from the first-path conventional farmers with whom I had previously worked. First, second-path farmers are much more diverse with respect to age, gender, education, and income. Second, more families, including children, attend sustainable agriculture conferences, and the whole family partici-pates, often as presenters and well as attendees. Third, these new farmers willingly share ideas and information; they are trying to help each other succeed. Perhaps because of the other differences, these second-path farmers tend to be much more hopeful, if not optimistic, about the future of farming than are their first-path counterparts.

There is a crisis in American agriculture today. The tradi-tional farming middle class is disappearing. But within this cri-sis, there is an exciting opportunity to create a new middle class of farming. I have struggled to understand the realities of an agriculture gone wrong. I have searched for logical reasons to hope for a brighter future for farming. And, I have reached the conclusion that the only realistic hope for the future of farming in America is along the second path—toward a more sustain-able agriculture.

I am not suggesting that all farms of the future will be small. However, I believe most will be smaller than the middle-class farms of today and virtually all will be smaller than are the larg-est farms of today. In sustainable farming, size, large or small, is not an objective; size is a consequence of farming sustain-ably—in harmony with community and nature. Sustainable farmers of the future will earn far more profit for each pound of product they produce, because they will put far more value into each pound they produce. Their farms won't need to be as large to achieve a middle-class farm income and an upper-class quality of life.

But I am not suggesting that middle-class farms of the future will be as small as most direct niche-marketing farms of today. The "agriculture of the middle" project is just one of many cur-rent public and private initiatives attempting to move sustain-able agriculture beyond the farm gate and into the food system by expanding opportunities for accessing higher-volume food markets. Many independent food processors and retailers are beginning to realize they face the same threats from a corporately controlled global food system as do independent family farms. They are becoming more open to forming alliances with groups of local farmers to create a new sustainable food system. Such alliances will provide more sustainably produced food to more caring customers, creating new opportunities for larger farm operations, as well as more opportunities for small farms. The critical challenge in accessing high-volume markets is to maintain the integrity of the system—not just the integrity of food quality and safety but also integrity of relationships—among eaters, retailers, processors, farmers, and through farmers, with the land. A new sustainable food system could be a key element in restoring the middle class of farming.

I am not trying to shove the term sustainable agriculture down anyone's throat. It really doesn't matter what you call it. If it's sustainable, it will contribute to a more desirable quality of life, and that's what's most important. If you don't like the ecological or sustainable labels, you can call it practical farming, balanced farming, true family farming, or commonsense farming. Perhaps the basic ideas would be clearer to today's middle-class farmers if we referred to it as farming for the triple bottom line.

The business concept of a triple bottom line first came to widespread attention in corporate management circles in the late 1990s and has since gained in popularity among businesses of all types.[10] Managing for a triple bottom line suggests managing for balance among the economic, environmental, and social dimensions of business performance rather than maximizing profits or growth. Triple-bottom-line managers recognize that businesses lacking social and ecological integrity are not economically viable over the long run; their costs eventually increase and the loyalty of their customers eventually declines. So they focus on conserving nonrenewable resources and protecting the environment by being a good neighbor and a responsible citizen, as a means of maintaining long-run profitability.

In many situations, they find that by paying more attention

to social and ecological performance, they can actually improve economic performance, even in the short run. They may find ways to transform wastes into economic inputs and to increase production while using fewer costly nonrenewable resources.[11] They may also find ways to reduce labor costs and create new markets by developing and maintaining better relationships with their workers, their customers, and others in the communities in which they operate.[12] In general, they improve their efficiency in converting ecological and social resources into economic advantages.

However, triple-bottom-line management has its legitimate skeptics. Businesses have always claimed to be good neighbors and good corporate citizens, but such claims have rarely been allowed to take precedence over maximizing corporate profits.[13] Even Monsanto and DuPont, for example, have "sustainable agriculture" programs. In such cases, the triple bottom line becomes little more than a public relations strategy. On the other hand, Ray Anderson of Interface Inc., one of the largest manufacturers of carpets, is a well-known exception to this strategy of deception. Anderson travels the country proclaiming the benefits of triple-bottom-line management and provides his corporate financial records as compelling evidence that even a large publicly owned corporation can be profitable as well as socially and ecologically responsible.[14] In the food business, Paul Dolan, former chief executive officer of Fetzer, the sixth largest winery in the United States, is a prime example of a triple-bottom-line manager.[15] New Seasons Market in Portland, Oregon, a locally owned, rapidly growing five-store modern supermarket chain managed by Brian Rohter, provides another example of a food business managed for the triple bottom line.[16]

So managing for sustainability is not just for small farmers and niche marketers, it is a successful business strategy for some of the most successful businesses in the United States. However, the true triple-bottom-line manager, large or small, must be willing to give as high a priority to being a good neighbor and being a good steward of nature as being a profitable business. At times, this means that profits will be less than if the manager had been willing to pollute a little more or exploit a little more to cut costs or increase sales. The true triple-bottom-line manager must realize that his or her advantage and uniqueness is in the integrity of the business—in its commitment to good citizenship and stewardship—not in short-run economic efficiency. If that integrity is ever compromised for the sake of economic efficiency, the uniqueness is lost and the market advantage is gone. True triple-bottom-line management requires a faith that valuing right relationships among people and with nature is the right strategy to succeed in business.

At first, though, a business strategy based on right relationships may seem a bit naive or idealistic, but on further thought, it is not. Our first thought may be that our highest priority should be on economics, but further thought will reveal that economics is only a means to a greater end, in business and in life. Historically, it was generally accepted that living was about the pursuit of happiness, not just the pursuit of wealth. Wealth, at most, was only a means to finding happiness. The Founding Fathers of the United States were so bold as to identify the pursuit of happiness among the inalienable rights of all people. In fact, it's only within the past century that economics has abandoned the pursuit of happiness for the pursuit of wealth.

Early-nineteenth-century economists, including such notables as Adam Smith and David Ricardo, considered happiness to be the ultimate goal of all economic activity. Smith wrote of self-interest, but he also wrote, "No society can surely be flourishing and happy, of which the far greater part of the members are poor and miserable."[17] Ricardo, the father of free-trade theory, defended trade as being important to the "happiness of mankind."[18] Neither assumed that greater wealth was synonymous with greater happiness.

However, at the turn of the twentieth century, Italian economist Vilfredo Pareto set about to free economics from the subjectivity of sociology and psychology by focusing on what he called "revealed preferences" rather than happiness.[19] Obviously, rational persons would make rational choices, thus revealing their preferences for the things they want and need. Economists should focus on consumer choices, he suggested, and let the sociologists and psychologists worry about whether such choices actually make people happier. Pareto's theories eventually were adopted by other economists, primarily because it allowed economics to focus on observable and measurable human behavior rather than on some intangible concept of human happiness.

In the early 1900s, another noted economist, Alfred Marshall, conceded that economics no longer dealt directly with human "well-being," his term for happiness, but rather with the "material requisites" of it.[20] Latter twentieth-century economists, including England's John Hicks and America's Paul Samuelson, however, made little distinction between wealth and happiness.[21] They needed objective, quantifiable economic variables to accommodate their mathematical and statistical models. Maximizing profit, income, or wealth became equivalent to maximizing satisfaction or happiness, as far as these neoclassical economists were concerned.

Regardless of what economists suggest, our common sense tells us that wealth does not bring happiness, because happiness requires more than having lots of money to buy lots of stuff. Happiness has been a widely discussed and debated issue among the world's greatest philosophers. The Hedonist philosophers equated happiness with sensory pleasures. That's what today's economics is about—short-run, individual, material self-interests. However, another group of philosophers, including Aristotle, used the term *eudaimonia* for happiness. Eudaimonia is inherently social in nature—it is realized by the individual but only within the context of family, friendships, community, and society.

Aristotle's happiness, social happiness, is a natural consequence of positive personal relationships—the individual in harmony with society. In addition, this social happiness was considered a byproduct of actions taken for their own sake—not to achieve some sensory satisfaction, but actions taken because they are intrinsically right and good. In essence, Aristotle and his followers believed that personal happiness was a natural consequence of right relationships. So it is not naive to believe that managing a business with integrity and pursuing right relationships through true triple-bottom-line management is the key to long-run business success. Valuing right relationships is the right strategy for success in business as well as in life.

If farmers are to find the courage to follow the second path to sustainability, to restore the middle class of farming, they must learn to rely on their own common sense of right and wrong. The agricultural establishment is not going to help farmers free themselves from dependence on costly farm inputs, machinery, and technologies. Agribusinesses have built their prosperity and agricultural professionals have built their careers on these things. The agricultural establishment will continue to promote the conventional wisdom of farming for the economic bottom line, even though common sense tells us that triple-bottom-line farms are the farms of the future.

The agricultural establishment tells the rest of us that farmers must maximize production to feed a hungry world. But our common sense tells us that hunger in the world today is not due to a lack of food, but to a lack of concern among those who *have* for those who *have not*. The establishment tells us we must invest in technologies today to feed 50 percent more people fifty years from now. But our common sense tells us an agriculture that is dependent on nonrenewable resources, particularly fossil energy, eventually will run out of resources and lose its ability to produce.

The agricultural establishment says that a sustainable agriculture would take too much land and that no land will be left for wild places. But our common sense tells us that industrialization, not sustainable farming, is the greatest threat to wild places, that sustainability won't require more land, just more imaginative, creative, trustworthy, caring farmers. What's wrong with creating opportunities for more family farmers rather than fewer? Our common sense tells us that an agriculture that puts a priority on people and on nature is more likely to take care of people and nature than is an agriculture that puts its priority on profits.

But why should we listen to our common sense instead of the conventional wisdom of economics? Today we are told we must base our decisions on "good science" rather than common sense. However, even many scientists fail to realize that good science must be rooted in common sense. Science is built upon foundational first principles, which are used to test the truth of knowledge or to prove whether something is true or false. First principles likewise must be used to test the morality of actions or to judge whether something is good or bad.[22] As nineteenth-century philosopher Thomas Reid wrote, "All knowledge and science must be built upon principles that are self-evident; and of such principles every man who has common sense is competent to judge."[23] First principles provide a starting point, and lacking a starting point, all logic and reasoning become circular, and thus, useless. For example, first principles of algebra, called axioms or laws, are the foundation for all mathematical proofs. One such axiom is that a times b equals b times a. We can't prove this equality; we just accept it. It may seem obvious, but that's the nature of first principles. First principles are common sense. Thus, good science must be rooted in common sense.

Conventional wisdom, on the other hand, is simply secondhand opinion regarding what is true and right, something passed from person to person and from generation to generation. Conventional wisdom is based on hearsay or experiments and observations, on imperfect reflections of an unobservable reality. Common sense instead reflects our direct and personal insights into the true nature of things. Our common sense reflects what we know to be true and good, because we sense it in the very deepest part of our being.

American agriculture is in crisis. Within this crisis are risks but also opportunities. Farmers of the middle today are confronted with two paths to the future of farming: the first path leads toward corporate industrialization and the second path leads to agricultural sustainability. Industrialization is driven by a continuing need for productivity and growth, by a single economic bottom line. Industrialization leaves room for fewer farmers, because farms must become larger to survive, and thus, the disappearance of agriculture of the middle. Sustainability requires ecological integrity, social responsibility, and economic viability, not one single bottom line, but three bottom lines. Sustainable farming makes room for more farmers, and thus, the opportunity to create a new middle class of farming.

The economic bottom line is supported by the conventional wisdom of the agricultural establishment, but the triple bottom line is supported by the common sense of thoughtful, caring people everywhere. Industrialization may lead to greater wealth for the few, but sustainability leads to greater happiness for the many.

I believe in the future of farming in America once again, because I believe American farmers will somehow find the courage to challenge today's conventional wisdom of farming for the single economic bottom line. I believe in the future of farming because I believe American farmers eventually will find the courage to follow their common sense and to farm sustainably by farming for the triple bottom line.

┊┊ Presented at "Changing Agricultural Landscapes," sponsored by the Northern Tier Cultural Alliance, Troy, Pennsylvania, September 30, 2004.

PART FIVE

Creating Sustainable Food and Farming Systems

17

The Real Costs of Globalization

During the 1990s globalization became an issue of broad public concern. Most of the controversy has centered on the World Trade Organization (WTO).[1] The WTO was established in 1994 to replace the General Agreement on Tariff and Trade (GATT) and was given authority to oversee international trade, administer free-trade agreements, and settle trade disputes among member nations. However, the authority of the WTO was greatly expanded to cover trade in services as well as merchandise and protection of intellectual property rights, including copyrights for artistic recordings and computer programs and patents for genetic materials. The WTO also was given far greater authority over trade in agricultural commodities than had existed under the GATT. The implicit, if not explicit, objective in forming the WTO was first to reduce and eventually to remove all obstacles to world trade, to achieve a single global free market.

Globalization is far broader in meaning than is "global free market." According to the dictionary, to "globalize" means "to make worldwide in scope or application."[2] The objective of the WTO is to create a marketplace that is worldwide in scope. However, we cannot globalize markets without also affecting the larger global economy, ecology, and society. This is the crux of the current WTO controversy. What are the real benefits and costs of globalizing markets, not just for the world economy but also for the world community and the world as a whole?

We live in a global ecosystem, the biosphere, whether we like it or not. We have no choice; such is the nature of *nature*. The atmosphere is global. Whatever we put in the air in one place eventually may find its way to any other place on the globe.

243

Weather is global. The warming or cooling of the oceans in one part of the world affects the weather in another, which in turn affects the temperature of oceans elsewhere on the globe. Thus, the oceans also are not just multinational waters, they are truly global waters. All the elements of the biosphere are interrelated and interconnected, including its human elements. We are all members of the global community of nature. We have no choice in this matter.

Increasingly, we are all living in a global community. Global communications—print media, radio, television, and the Internet—have erased national communications boundaries, resulting in the spread of common social and cultural values around the globe. Global travel has become faster, easier, and less expensive, resulting in greater person-to-person sharing of social and cultural values among nations. Consequently, the distinctiveness of national cultures has diminished. We seem to be moving toward universal membership in a common global culture.

However, in matters of culture we have the right and the responsibility to choose. We have the right to maintain whatever aspects of our unique local or national cultures that we choose not to lose. And we have the responsibility to protect this right against the economic or political forces pushing us toward a single global culture.

We have also been moving toward a single global economy. International trade increased dramatically during the 1970s, 1980s, and 1990s, first under the various GATT agreements and then under the WTO. All of the national economies of the world are now interconnected through their dependence on each other for trade. Problems anywhere in the world economic community may create economic problems for nations all around the globe. However, removing all barriers to trade would greatly increase international dependencies by creating a single global marketplace.

In this matter we also have a right and a responsibility to

choose. Every nation has the right to maintain those aspects of its local and national economies that are necessary to protect its resources and its people from exploitation. With a single global marketplace, the social and political boundaries that now constrain economic exploitation would no longer exist. The effect would be to move beyond economic globalization to economic homogenization. The people of every nation have a responsibility to decide whether they want to participate in this economic homogenization. Again, the crux of the WTO controversy is in weighing the benefits and costs of removing the economic boundaries that now stand in the way of a single global marketplace.

Perhaps the best way to begin addressing this question is to ask why boundaries exist in the first place. The boundaries in nature—ecological boundaries—are there by nature. Natural features, such as oceans, mountains, and even rivers and ridges, separate one physical bioregion from another. Why do we find such boundaries in nature? Perhaps boundaries are nature's way of maintaining its diversity. Boundaries define the natural form or structure of things that support life: sunlight, air, water, and soil. Boundaries also define the structure of living things: bacteria, fungi, plants, animals, and humans. We know that diversity of life is necessary to ensure the resistance, resilience, and regenerative capacity to healthy living ecosystems. Without diversity, without boundaries, nature could not support life, including human life.

Cultural and political boundaries are those that define communities of people—including cities, states, and nations. People have established such boundaries to facilitate relationships among members of communities and to differentiate relationships *within* communities from relationships *between* communities. Within community boundaries, relationships have been formed and nurtured to enhance social connectedness and personal security. Boundaries *between* communities have maintained a sense of community identity and thus have maintained cultural and social diversity. People have valued diversity as a means of maintaining choices and opportunity, which historically have been deemed necessary for the health, growth, resilience, and long-run security of human society.

In earlier times, cultural and political boundaries tended to coincide with natural boundaries—oceans, mountains, rivers, and ridges. During the industrial era, however, there was a tendency to ignore the guidance of nature, to allow economic and political considerations to take priority over nature in defining political boundaries. Wars have redrawn boundaries of countries along lines that have little relationship to either topography or culture. Towns and cities have expanded their boundaries with little regard for the best long-run use of the land now covered with buildings and concrete. And with the trend toward a single global community, the remaining social and cultural boundaries that once defined distinct groups of people with different social, ethical, and moral values are being largely ignored.

With some notable exceptions, economic boundaries have been the same as national political boundaries, at least over the past century. Historically, each nation has had its own rules of trade, defining economic relationships among people *within* nations, which were distinctively different from the international rules of trade, which defined economic relationships *among* nations. The British Empire of the early 1900s, which once included about a fifth of the globe, might have been considered a single economic unit. More recently, the North American Free Trade Agreement (NAFTA) and the European Union (EU) represent attempts to bring several nations within a single economic boundary.[3] But most economic communities have been defined historically as single nations.

The purpose of economic boundaries is to promote free trade within boundaries and carry out selective trade across boundaries. Economic diversity among nations has been considered a necessary means of ensuring choice and opportunity, which historically has been deemed necessary for health, growth, resilience, and long run security of the global economy. Humanity simply has not been willing to put all its "economic eggs in one basket."

So why are leaders of the major economic powers of the world now seemingly willing to put all their economic eggs in the WTO basket? The most logical answer seems to be that world leaders are now motivated more by short-run economic consideration than by longer-run concerns for either human culture or the natural environment. In this respect, other nations quite likely are being misled by the economic culture of the United States, which dominates the global economy. The tremendous growth of the U.S. economy over the past century is widely attributed to our so-called competitive free-market economy. Admittedly, this new culture of economics now holds sway among many in the most economically powerful nations of the world.

Within this culture, economic boundaries are viewed as obstacles to trade, which limit the ability of investors to maximize economic efficiency and thus limit economic growth. Free trade among all nations would result in a more efficient global economy, thus benefiting all people of the world. In the culture of economics, barriers to trade are seen as nothing more than artificial political constraints designed to protect specific individuals and industries within nations from economic competition with more efficient producers in other nations. Thus the WTO is working to remove such barriers, to allow the most efficient producers in the world to produce the world's goods and services, resulting in lower cost to consumers everywhere.

Such beliefs are based on economic theories of trade that historically have made free trade something of a sacred tenet of economics, particularly among the more conservative of economists who are now in vogue. Modern free-trade theory has its foundation in the early 1800s writings of British economist David Ricardo.[4] Ricardo explained that when two individuals choose to trade, each is better off after the trade than before the trade. People have different tastes and preferences and thus different people value the same things differently. So if I value something you now own more highly than I value something I own, and you value the thing that I own more highly than you value the thing you own, we will both gain by trading. I get something I value more than the thing I now own and so do you.

The same concept can be used to show the potential gains from trade associated with specialization. One farmer may be a more efficient producer of one thing, say corn, and another farmer may be a more efficient producer of another, say cattle. If so, both can be made better off if one farmer specializes in cattle and the other in corn. The better corn producer can then trade corn for beef and the cattle producer can trade beef for corn, and they both will be better off than if they each tried to produce both beef and corn.

Even if one farmer is a better producer of both beef and corn, the other farmer will have a "comparative advantage" in producing one or the other. Let's say the first farmer could produce either a 1,200-pound steer or 300 bushels of corn with a given amount of land, labor, and capital. Assume the second farmer could produce only a 750-pound steer or 250 bushels of corn using the same amount of resources—not as much of either—as could the first farmer.

If the first farmer decided to produce only corn, he or she would have to forgo four pounds of beef for each bushel of corn produced (1,200 divided by 300). However, if the second farmer decided to produce corn, he or she would have to forgo only three pounds of beef for each bushel of corn (750 divided by 250). In economic terms, this means that the second farmer has a "comparative advantage" in producing corn because his or her "opportunity cost" of producing corn is less. The two farmers will have to forgo less beef for each bushel of corn if the second farmer uses his or her land, labor, and capital to produce corn and the first farmer produces the beef. Using the same logic, the first farmer has a lower "opportunity cost" of producing beef—one-quarter bushel of corn per pound of beef (300/1200) compared with one-third bushel per pound (250/750) for the second farmer.

Although the arithmetic gets messy, if the second farmer specializes in corn and the first in beef, and they trade their surpluses with each other, both will be better off than if each produces their own corn and beef. Of course the real world is much more complex than this simple two-farmer, two-commodity example, but this simple one-on-one trade situation is still at the heart of neoclassical economic trade theory.

So if both traders gain from specialization and trade, what's wrong with free trade? The problems arise because free trade between two independent individuals, in the context of the early 1800s, does not accurately reflect the reality of trade among nations in the early 2000s.

First, trade is truly free only if both partners are free not to trade. Participants in free trade must have an interdependent relationship, meaning that they depend on each other by choice, not by necessity. If one trading partner is dependent on another, the dependent partner may have no choice but to do whatever is necessary to maintain the relationship. When both are independent, neither has any obligation to maintain the relationship. Interdependent relationships can only be formed between two independent entities. Under such circumstances, relationships are formed only if they are beneficial to both and continue to exist only so long as they remain beneficial to both.

Under the WTO, however, stronger nations are granted the ability to force weaker nations to form dependent trading relationships, creating situations where weaker nations are not free to *not* trade. Trade made under conditions of coercion, under explicit or implied threats of retribution if one does not trade, is not free trade. The school kid who trades his lunch to one bully in return for protection from another bully is not participating in free trade. Neither is a weak nation that trades with a strong nation under the threat of denial of military protection from some global tyrant. Nor is it free trade if one nation is dependent on the other for its economic well-being, as in cases where one nation has large debts to be repaid to another. Poor nations are made dependent on rich nations by their lack of economic wealth, economic infrastructure, and technological advantage, regardless of their inherent worth to humanity. In many cases, rich nations are able to exploit the workers and resources of poor nations through trade. The poor see no other way to avoid physical deprivation or starvation of their people. Coerced trade is not free trade.

Second, free trade assumes informed trade. Both parties must understand the consequences of their actions. If a car dealer trades cars with a customer, knowing that the car is a gas guzzler, needs lots of repairs, and is unsafe to drive, but without informing the customer, this is not a free trade. When a developed nation encourages a lesser-developed nation to produce for export markets, knowing that such production will lead to exploitation of their natural and human resources, and does so without informing them of the consequences, this is not free trade. The leaders of the lesser-developed nations may benefit from such trade, perhaps from bribes or payoffs from the outside exploiters, but the resources of the lesser-developed nation will be exploited rather than developed. The people will be left with fewer opportunities for developing their country than before. The exploiters know the consequences but the exploited do not. Uninformed trade is not free trade.

Third, free trade, in economic theory, implies that the decision is made by an individual, not a nation. Individuals are whole people, presumably absent of unresolved internal conflicts regarding the relative values of items being traded. A person trades only if he or she decides that the trade, overall, is good for him or her. Nations, on the other hand, may make and carry out trade agreements to which a substantial portion, perhaps a majority, of the nation's population is opposed, both before and after trade takes place. The economic rationale for such agreements is that if the economic benefits to those who favor trade more than offset the economic costs to those opposed, the nation as a whole will benefit from trade.

Economics is incapable of dealing with issues of equity and justice. In economics, a nation is said to gain from trade if those who benefit from trade could compensate those who lose and still have something left over. Of course, the gainers have no legal obligation to compensate the losers and rarely, if ever, do so. From a purely economic perspective, it doesn't matter that the rich are made richer and the poor are made poorer. It doesn't matter how many people are made relatively worse or better off by trade, as long as the trade results in growth of the economy. In neoclassical economics, trade among nations is no different in concept from trade among individuals.

Finally, the foundational principles of economic trade theory are rooted in a "barter economy," in which one person trades something to another. In an international currency-based economy, comparative advantages in trade often are distorted by fluctuations in exchange rates resulting from differences in monetary policies among nations, which may have no relationship to relative productivity. Such fluctuations can cause the exports from one nation to become more or less costly to importers from another nation for reasons totally unrelated to differences in production efficiency. Under such conditions, free markets do not result in efficient use of resources.

Also, in classic trade theory each trading partner uses his or her individual resources—land, labor, capital, and technology—to do whatever he or she does best—to exploit comparative advantages. No consideration is given to the possibility that one nation might instead choose to transfer some of their resources, such as capital and production technology, to another nation as a means of generating greater profits than are available through trade. Mobility of capital and technology, hallmarks of today's global economy, eliminates the "comparative advantage" of higher-cost nations, forcing them to import from lower-cost nations and devaluing both land and labor in the higher-cost nation to lower, globally competitive levels.

Because of these inconsistencies between economic theory and economic reality, most international trade today does not fit the classical theory of economic free trade. Perhaps more important, widespread opposition to and open defiance of the WTO by countries all around the globe indicate that any future expansion of trade that is forced on people by the WTO almost certainly will not be free trade. The protesters believe, with good cause, that the current WTO version of free trade actually is coerced, uninformed trade, resulting in the exploitation of the weak and poor by the strong and wealthy.

If the free-market goals of the WTO were achieved, all national restraints to trade would be removed. Initially, all market boundaries would be translated into tariffs, and over time all tariffs would be eliminated, erasing all market boundaries among nations. The global market would presumably operate pretty much as national markets operate today. International commerce would be a lot like interstate commerce, as no nation would be allowed to enforce laws interfering with such commerce. The WTO, not individual nations, would decide what products and resources nations could and could not exclude from international commerce. If the WTO decides that nations have to open their national parks and historic sites to mining and oil exploration, nations must lease such land to the highest global bidder. If the WTO decides that clean water and clean air are marketable commodities, pristine nations will be forced to open their borders to global pollution. In addition, no seller or buyer would be allowed to offer a different price or conditions of trade to buyers of one nation, including one's own nation, than it offers to buyers of any other nation.

Under such rules of trade, a nation could not subsidize its

agriculture by any means that might distort their comparative advantage in trade; that is, a nation couldn't subsidize producers of one commodity more than it subsidizes producers of another. A nation could not establish environmental, health, or safety standards for its production processes that were more restrictive than those specified by the WTO. A nation could not close its borders to WTO-approved cultural exports from other nations—movies, television programs, clothes, and magazines—no matter how culturally repulsive they may be to current residents of that nation. A nation could not refuse to sell its natural resources, such as minerals, oil, or even water, to another nation. And, the WTO would stand ready to enforce merchandise patents and intellectual property rights globally, regardless of whether the people of the world agree that all things, such as life forms, should be patented.

One might expect U.S. farmers to benefit from the WTO, even if farmers of other countries pay the costs. After all, the United States is the strongest of the strong nations and the strongest promoter of the free-market goals of the WTO. But what are the *real* costs of globalization to American farmers?

Until a decade or so ago, few questioned the ability of American farmers to compete with farmers anywhere in the world. All they needed was a "level playing field." We were the self-declared global leaders in agriculture. We had the most highly educated and efficient farmers in the world using the latest production technologies to cultivate the best agricultural land in the world. However, during the 1990s, the U.S. share of global agricultural exports plummeted, dropping farm profits and shaking confidence in the American farmer's ability to compete.

Rising costs of land and labor have destroyed the traditional competitive advantage of American farmers in world markets, and costs of neither land nor labor can drop to globally competitive levels without dragging the American economy along with them. American farmers traditionally have relied on superior access to capital and technology to offset any disadvantage in resource costs, but global agribusinesses eventually will utilize their capital and technologies in places with lower-cost land and labor in order to maximize returns to their stockholders.

Under comprehensive corporate contracts, corporations, not farmers, select the genetics of crops and livestock, decide how the growing crops or animals are to be managed, and make all the important decisions, including where and with whom they choose to contract. These same corporations eventually will control access to global commodity markets, and American producers without contracts will not have access to those markets. These multinational agribusiness corporations have no sentimental ties to family, community, or even to any given nation, because they are not real people and their stockholders may be located anywhere on the globe. They eventually will move their agricultural operations, including contractual operations, to wherever on the globe they can make the most money, which is not likely to be the United States.

So what will be the real cost of globalization to American farmers? Perhaps the cost will be the lost opportunity to farm, at least farm in the sense that we have known it in the past. The end of the American farm could well be one of the real costs of globalization.

Economists argue that it doesn't matter where our food is produced. If producing it elsewhere in the world will be cheaper, we will all be better off without agriculture in the United States, so they say. But how long will it be before an "Organization of Food-Exporting Countries" is formed to restrict world food supplies, causing our food prices to skyrocket—as we have seen OPEC do with our energy prices in the past. Even more importantly, we have only a few days' supply of food in the "food pipeline" at any point in time. The disruption of global food supplies, even for a short period of time, could have devastating consequences for millions of people.

Perhaps we could keep our food imports flowing, through

our military might, if economic coercion fails. But what will be the real costs? How many more terrorist attacks might we expect as a result of our global food policy? How many small wars will we feel compelled to fight? How many people will be killed to support a global food system? The highest *real* costs of globalization may be paid in human blood.

Even if the United States somehow maintains its food security, nations with less productive resources are almost certain to become subject to nutritional blackmail in the new global economy. For example, if African nations want access to food, they must open their borders to genetically modified organisms to which global corporations hold exclusive patent rights. Those nations with more than enough food inevitably will threaten to stop selling their surpluses to those who don't have enough, as the United States has withheld food from our so-called enemies in the past. Even more importantly, those corporations controlling global food production in the future will use their newfound political power to shape the policies of every nation of the world, including the United States.

With the multinational corporations in control of the global food supply, the resources of no nation will be secure from exploitation. There will be no effective limits to their ability to exploit, pollute, and destroy. And almost certainly, with corporate control of the food economy, food prices are far more likely to rise than fall. And those without the means of paying higher prices for food will be more likely to starve. A major cost of globalization may be the loss of food security for people of both rich and poor nations of the world.

Finally, what are the costs of globalization of the food system to global society? The answer is, quite possibly, the sustainability of human life on earth. The question of sustainability is "How can we meet the needs of all people of the present, while leaving equal or better opportunities for those of the future?" The answer is "Through systems of production and distribution that are ecologically sound, economically viable, and socially responsible." Globalization is a strategy designed for short-run economic exploitation, not long-run societal sustainability.

To be ecologically sound, a sustainable food system must work in harmony with nature, not attempt to dominate or conquer nature. Nature is inherently diverse. Diversity in nature is necessary to support life within nature. Boundaries in nature define the diversity of landscapes, life forms, and resources needed to support healthy, natural, sustainable production processes. Fencerows, streams, and ridges define unique agroecosystems within which nature can sustain different types of human enterprises. Globalization will remove the fencerows, divert the streams, and level the ridges to facilitate standardization and homogenization of production processes. The natural boundaries needed for sustainability will be removed to achieve greater economic efficiency. A *real* cost of globalization to humanity will include the loss of ecological sustainability.

To be socially responsible, a sustainable food system must function in harmony with human communities, including towns, cities, and nations. Humanity is inherently diverse. Diversity among people is necessary for interdependent relationships—relationships of choice among unique, independent individuals. Although we have our humanity in common, each person is unique, and we need unique human communities within which to express our uniqueness. Social and cultural boundaries define those communities—including towns, states, and nations. Economic globalization will remove those boundaries and will homogenize global culture and society. The natural boundaries needed to sustain social responsibility will be removed to achieve greater economic efficiency. A *real* cost of globalization to humanity will include the loss of social sustainability.

To be economically viable, a sustainable food system must facilitate harmonious relationships among people and between people and their natural environment. The inherent diversity of nature and of humanity must be reflected in diversity of the economy. Although potential gains from specialization are real, such gains are based on the premise that people and resources are inherently diverse, with unique abilities to contribute to the economy. Competitive capitalism is based on the premise that individual entrepreneurs make individual decisions and accept individual responsibility for their actions. If globalization is allowed to destroy the boundaries that define the diversity of nature and people, then it will destroy both the efficiency and sustainability of the economy. A *real* cost of globalization to humanity will include the loss of economic viability.

The real costs of globalization quite simply are too high to pay. But what can we do to avoid paying these costs? How can we stop globalization? First, we can help people realize that the undeniable existence of a global ecosystem, global society, and global economy does not justify economic homogenization or the removal of all economic boundaries. Natural boundaries are necessary to ensure ecological integrity. Cultural boundaries are necessary to ensure social responsibility. And economic boundaries are necessary to ensure long-run economic viability. Without boundaries, the biosphere would be left without form, without structure, without order, and without life.

Nations undeniably can benefit from free trade, but it must be uncoerced, informed trade among sovereign entities. To benefit from free trade, nations must be free to trade or not trade. Each nation must be afforded its right and must accept its responsibility to protect its people and its resources from exploitation, just as all persons have a right and responsibility to protect themselves and their property from exploitation. A single global free market would deny these most fundamental of human rights to the communities of people that constitute the nations of the world.

People benefit from healthy relationships with each other and with the earth, but healthy relationships are relationships of choice. Global society needs a world forum—such as the wTO perhaps *could* be—not to remove boundaries, but to ensure that every person of every nation is protected from economic exploitation. To avoid the high costs of globalization, we must reclaim our rights to individual and national sovereignty.

Other things we can do to fight economic homogenization are more tangible and practical. For example, we can all help develop more sustainable local alternatives. Thousands of farmers and their consumers all across North America are already joining forces to develop more sustainable local food systems. These people come together regularly within local communities, at farmers' markets, community gardens, and other venues where farmers and consumers meet around food. In addition, the increasing number of large conferences, bringing farmers and consumers together around common concerns for food safety, nutrition, environmental quality, social justice, and other issues of sustainability, indicate a growing interest in local food systems. Farmers can give priority to local markets in developing more sustainable farming systems. The rest of us can buy as much of our food as possible from local farmers. We can all help to develop a local sustainable alternative to globalization.

Supporting local food systems doesn't mean that we have to give up oranges, bananas, coffee, or things that can't be produced locally. Trading when we are free not to trade can be beneficial to all concerned. We simply need to buy and sell locally to the extent necessary to maintain the sustainability of our local food system. We can and should continue trading with those in other regions and other nations to help ensure the sustainability of agriculture everywhere on the globe. It's just that relationships among regions and nations must be interdependent rather than dependent, if the global food system is to be sustainable.

It would be easy to be skeptical about the possibility of success of local community-based food systems because such systems currently make up such a small part of the huge global food system. Farmers and consumers may seem too few and too weak to confront the giant global food corporations. However, the trend toward a global food system over the past several decades took place one farmer and one customer at a time. One by one, as farmers changed what they produced and where they sold their products, and as consumers changed what they ate and where they bought their food, a food system that had been local became global. Again, one by one, we can and must make the changes needed to create a sustainable local food system. Will we succeed in avoiding the high costs of globalization? I don't know if we will, but I know it is possible, and thus I have hope. Hope is not the expectation that something good is destined to happen or even that the odds favor something good, but rather, that something good is possible. I know that something better than globalization is possible. It is the very real possibility of a sustainable network of local food systems that gives farmers and consumers the courage to challenge globalization, with everything from protesting in the streets to buying and selling locally. Regardless of whether we ultimately win or lose in this struggle, life is simply too precious to live without hope. We simply can't afford the greatest possible cost of globalization; we simply cannot afford to lose hope.

⁞ Presented at the Eleventh Annual Sustainable Farming Association of Minnesota Conference, "Sustaining Our Food System: Creative Alternatives to Globalization," St. Olaf College, Northfield, Minnesota, February 23, 2002.

18

Redirecting Government Policies for Agricultural Sustainability

On signing the Farm Security and Rural Investment Act of 2002, President George W. Bush said, "The farm bill will strengthen the farm economy over the long term. It helps farmer independence, and preserves the farm way of life for generations. It helps America's farmers, and therefore it helps America."[1] Similar claims have been made for every U.S. farm bill since the 1930s. Yet the farm economy has continued to flounder, American agriculture has limped from one crisis to the next, and fewer family farmers have survived to witness the signing of each new farm bill. It's difficult to believe this farm bill will be any different.

Since the mid-1990s, prices for all major agricultural commodities—including corn, soybeans, wheat, hogs, and cattle—have averaged well below break-even levels for most farmers. Without U.S. government farm programs, now among the most costly in the world, the economic situation in American agriculture in the early 2000s would be as dire as during the farm financial crisis of the mid-1980s. Even if commodity prices rebound due to adverse weather conditions elsewhere in the world or some other market aberration, the fundamental problems of American agriculture will remain unsolved. Current farm policies will do nothing to address these problems.

Contrary to popular belief, livestock producers are no less reliant on government subsidies than are crop producers. Even without the various so-called emergency programs, Export Enhancement Programs (EEP), Environmental Quality Incentive Payments (EQIP), and other direct payments, livestock producers still would be dependent on government programs. The animal agriculture industry in the United States is built on a foundation of abundant supplies and stable prices for feed grains. For example, when corn farmers get subsidy payments from taxpayers rather than payments from the marketplace, corn prices are lower and livestock feeders pay less for grain, allowing them also to pay more to the producers of the animals they feed. The division of government benefits among corn producers, livestock feeders, and producers of feeder animals depends on their relative power in the marketplace, but all rely on government benefits.

Direct government payments to farmers averaged about $20 billion per year between 1999 and 2001, before declining to the $15 billion range for 2002 and 2003.[2] While government payments have generally averaged only 8 to 10 percent of *gross* farm income, government payments have accounted for about half of farmers' *net* farm income in the late 1990s and early 2000s. However, these government programs have done nothing to address the root causes of the current crisis in agriculture. Most government programs allocate payments in proportion to production or acreage. Thus, current programs mostly benefit corporate agribusiness and wealthy landowners to the detriment of the average family farmer, for whom government payments do little more than provide enough cash to farm or ranch another year.

The previous farm bill, called "Freedom to Farm," was supposed to provide a transition period designed to "get the government out of agriculture." American farmers and ranchers were to be allowed to compete freely in a new global agricultural economy. The Freedom to Farm bill removed most previous restrictions on production of agricultural commodities; farmers were free to plant as much as they wanted of virtually any crop they wanted to grow. Government payments were continued only as a "one-time incentive" for farmers to give up their reliance on government programs. Payments were based on "historical production levels" and were to be phased out over the five-year life of the bill. Growth in agricultural exports was to bring new prosperity to American farmers, making government price and income supports unnecessary.

However, freedom to farm soon became known as freedom to fail. U.S. farmers, feeders, and ranchers found that they simply couldn't survive at market prices offered by the new global economy. The U.S. share of world exports dropped for commodity after commodity—in spite of USDA's persistent forecasts that "farm exports are expected to improve next year." "Emergency" government payments, ranging from $4 billion to $9 billion per year, were approved by Congress to stave off widespread farm bankruptcies.

U.S. crop producers found that they couldn't compete with agricultural production from South America, Mexico, China, and other places where costs of land and labor are a fraction of costs in the United States. Livestock and poultry producers have fared little better in competing for world markets against livestock and meat products from Canada, Australia, and New Zealand. In addition, some major consuming countries, such as China and India, have found ways to reduce their reliance on agricultural imports. Perhaps even more troublesome is the persistent rise in U.S. imports of agricultural products from all around the world. The May 2003 USDA report "Outlook for U.S. Agricultural Trade" forecasts an export surplus of only $10.5 billion for 2003, the smallest surplus since 1987—in spite of an export forecast that appears even more optimistic than usual.[3]

If recent trends continue, the United States seems destined to become a "net importer" of food in the not too distant future. American farmers are losing their ability to compete in the global marketplace, even though the farm policy "playing field remains tipped in their favor." In a book widely read in agricultural circles, *The End of Agriculture in the American Portfolio*, University of California agricultural economist Steven Blank documents reasons for the growing comparative disadvantages of the American farmer.[4]

Historically, American farmers, feeders, and ranchers had been able to offset their global disadvantages of higher land and labor costs through greater access to capital and technology. However, these advantages no longer exist. In the new global economy, capital can and does move freely and quickly around the world, to wherever it can earn the greatest return on investment. The multinational agribusiness firms now control much of the new technology, biotechnology being a case in point. They are applying these technologies wherever in the world they expect to earn the highest return, increasingly somewhere other than in America.

America may still have the most knowledgeable farmers in the world, but knowledge—at least farming and ranching knowledge—is becoming less important in agricultural commodity production. Many of the new technologies have taken the unique farming and husbandry skills out of agricultural production. For example, Roundup Ready soybeans and confinement animal feeding operations (CAFOS) make it possible for virtually anyone to become a good contract producer. When farming and feeding can be done "by recipe," little real knowledge of agriculture is necessary, and agricultural production can be carried out by virtually anyone anywhere in the world.

The bottom line is that American farmers, feeders, and ranchers are no longer competitive in world markets and the new farm bill will do nothing to improve their competitiveness.[5] Modest changes in commodity programs, such as a virtual removal of all limitations on the size of payments, will subsidize large corporate operations to an even greater extent in relation to smaller family farms. In addition, funds appropriated for programs supported by environmental and family-farming advocates have been diverted in the rule-making process to subsidize agribusiness interests instead. Ten percent of agricultural producers, including many large corporate operations, received more than 60 percent of all commodity payments under the last farm bill.[6] Fewer large operations will get an even larger share under the farm bill of 2002.

The new farm bill does include some rays of hope for change, but they are difficult to see. The new Conservation Security Program promised to provide payments to any and all farmers who agree to be good stewards of the land and the natural environment. However, the devil may be in the details. In this case, the biggest question is whether Congress will ever allow the bill to take on "entitlement" status, making its benefits available to all farmers who qualify, or whether it will limit benefits to token year-to-year funding. The new farm bill also gives increased recognition of the legitimacy of organic farming, but programs supporting organic farming and all other approaches to sustainable agriculture combined still will receive less than 1 percent of public funding for research and education.

For every step forward there seems to have been two steps backward. Changes in the Environmental Quality Incentive Program (EQIP) now promises huge government subsidies to corporate confinement animal feeding operations, subsidizing the almost-certain continued destruction of the rural environment. A large portion of the funds appropriated for the Value-Added Marketing Program has been allocated to subsidize construction of ethanol plants and similar large-scale ventures, which almost certainly will end up in the hands of large corporate ethanol suppliers, such as Archer Daniels Midland. A number of Senate proposals that would have helped to limit corporate market power by restoring competition to agricultural markets were defeated either on the Senate floor or in the Conference Committee. And, agribusiness corporations were left with a virtual free rein to force farmers into signing comprehensive production contracts as their only means of maintaining access to markets.

All of the dominant players in the agricultural policy–making process have vested interests in maintaining high levels of production. Profits of agribusiness corporations depend on margin and volume, not on farm-level price. Surplus production means a higher demand for marketing services, resulting in wider profit margins on larger volumes of sales for marketing firms, even if farmers lose money. Surplus production also means more sales of seed, fertilizer, pesticides, and other production inputs and thus more profits for input suppliers, even if farmers can't cover their total costs. And with a limited domestic demand, increased production translates into high levels of agricultural exports, which are possible only if commodity prices are kept competitive, meaning *low*, even if American farmers continue to lose money.

Commodity organizations apparently want to keep production levels high because most are funded by checkoff programs that assess producers a given amount per head, per hundredweight, or per bushel of production. Those that aren't funded by checkoffs still appear to put the status of their commodity ahead of the profitability of their farmers, because most farmers produce several different commodities. Agricultural specialists at USDA and in the land-grant universities tend to share a similar mentality. They want to maintain the importance of the particular commodity in which they specialize and to maintain the importance of agriculture within the national economy. Increased production tends to be translated into increased importance. So they promote production of their chosen commodity, even if farmers who produce it are losing money.

However, the farm commodity organizations, particularly national beef and pork producers' associations and the Farm Bureau Federation, have come under increasing criticism from their rank-and-file farmer members as their true allegiances have become more widely known. Many farmers have come to view USDA and the land-grant universities with increasing skepticism because of their close financial and professional alliances with corporate agribusiness, particularly on biotechnology issues. American farmers are just beginning to understand that the future of farming and the future of the agricultural industry are two distinctively different concepts.

Ultimately, the food security of the United States depends on the productivity of its agricultural land and on the viability of its independently operated family farms and ranches that care for the land. As Kentucky farmer and writer Wendell Berry articulates so well, "If the land is to be used well, the people who use it must know it well, must be highly motivated to use it well, must know how to use it well, must have time to use it well, and must be able to afford to use it well."[7] The food security of the United States depends on farmers and ranchers on the land who know and love their land. Consequently, if farm policy is to ensure the long-run food security of the nation, it must ensure that these independent farmers and ranchers who use the land have the time to use it well and are able to afford to use it well.

Large corporate producers have no commitment to any particular piece of land; most don't even own most of the land they farm. They can't really know the land because they are trying to farm too much of it to know any of it very well. Many don't know how to take care of the land; they depend on a prescribed regimen of commercial inputs for their productivity, not on healthy soil. They can't really afford to love the land because they must stay focused on the bottom line; they have to stay competitive. American food security depends on having more, smaller, independent family farmers. A farmer or rancher can know only so much land, and thus can truly love only so much land.

Thankfully, a new type of agriculture is emerging to address the current crisis in American agriculture. Groups of creative, innovative, entrepreneurial farmers all across the country are seizing the opportunities inherent within the necessity for change—they are creating the New American Farm.[8] These new farmers and ranchers are given a variety of different labels by different people, but they are all pursuing the same basic purpose by the same set of principles. These New American Farmers are creating new systems of farming that take care of the land and help build strong communities while providing a good quality of life for their families. They are discovering ways of farming that are ecologically sound, economically viable, socially responsible, and thus will be sustainable over time. Almost invariably, these new sustainable farms are smaller, independently owned and operated family farms. By redirecting farm policy toward ensuring the economic viability of these smaller, independently operated family farms, we can go a long way toward ensuring our long-run food security.

Ironically, much of the current public support for government programs for agriculture stems from the belief that today's programs are already targeted to helping smaller, independent family farmers and ranchers. There is almost nothing to support this belief. Government payments may have helped farmers put in another crop during times of economic stress, but they have done nothing to secure the economic future of family farms and ranches. It's absurd to argue that current farm policies ensure either farm or food security while those policies subsidize the corporate industrial systems of production that are forcing farmers to become contract producers and thus placing our food security at risk. Fortunately, more and more people each year are being made aware that current farm programs are not working for the good of farmers, consumers, or the general public. This growing public awareness creates an opportunity for change.

Congress eventually must find the courage to focus agricultural programs on using public funds to produce public benefits—not for private subsidies. The societal benefits of agriculture, such as food security, are benefits that accrue to the public—to the people in general. The ecological benefits of agriculture, such as protection of water quality, accrue to the public—not to specific individuals or corporations. The creation of public benefits must become the focus of all publicly funded farm programs.

The private economy provides food and fiber for those who are able to pay the cost. And the prospects of profits provide adequate incentives for investments in the private food and fiber economy. But private markets will not provide adequate incentives for investments needed to ensure the social and ecological benefits from agriculture. Thus, we must make social and ecological investments collectively through government. If the potential ecological and social benefits of agriculture are to be realized, they must be encouraged through public rather than private investment—through government programs.

This is not a radical concept. For several years the Europeans have argued that agriculture is "multifunctional" in that it performs social and ecological functions in addition to its private economic functions. This has been their consistent position in world trade negotiations. Many Europeans understand the consequences of food insecurity because they remember World War II. The Europeans have argued that each nation should be allowed to maintain government programs necessary to ensure long-run food security. They have a deeper appreciation of the public benefits of having larger numbers of farmers on smaller farms in order to take care of the land and to support rural communities. They have argued that reducing trade restrictions on private markets should not preclude governments from ensuring that the public continues to benefit from the multifunctional aspects of agriculture. It is not radical to claim that governments have both the right and the responsibility to protect their people and their natural resources from economic exploitation.

The cornerstone of a new American farm policy should be long-run food security through agricultural sustainability. A sustainable agriculture must be ecologically sound, economically viable, and socially responsible. An ecologically sound agriculture provides clear benefits to the public, both now and in the future, beyond the economic benefits to farmers. A socially responsible agriculture provides clear benefits to the public, both rural and urban, beyond the economic benefits to farmers. An economically viable agriculture provides clear benefits to the public in terms of long-run food security, beyond the short-run economic benefits to individual farmers.

A government farm program based on long-run sustainability would be fundamentally different from the Farm Security and Rural Investment Act of 2002. First, with respect to ecological integrity, government farm programs eventually must recognize that no one has the right to degrade the natural environment. Thus, all farmers and ranchers should be required to meet environmental standards that conserve the soil, protect the quality of water and air, and in general, ensure the integrity of the natural resource base. The rights of private property have never included a right to destroy the productivity of the land or to degrade the natural environment. New ecological programs, such as the Conservation Security Program, should limit payments to rewarding farmers who rebuild soil fertility, restore water quality, and enhance the natural environment.

A socially responsible agriculture must provide farmers and ranchers, as people, with opportunities to lead productive and successful lives. This doesn't mean that all who choose to farm or ranch have a right to do so, regardless of their aptitudes or abilities. However, those who choose to farm or ranch, and are willing and able to farm or ranch sustainably, should be given an opportunity to do so. To support such opportunities, government benefits should be limited to individually owned and family-operated farms and ranches, and the benefits should be paid only to real people, not to corporations. The objective should be to provide self-employment opportunities for farmers, ranchers, and others in rural areas, not to subsidize landowners and corporations. The overall goal of any new American farm policy should be to keep enough independent family farmers and ranchers on the land who are committed to farming and ranching sustainably, to ensure the long-run food security of the nation.

The same dollars used to support current farm programs would be more than adequate to fund the new long-run food security program. And in contrast to existing farm programs, a sustainability-based farm program could be designed to be self-liquidating over time. In addition, the administration of such a program could be far less complex for farmer and rancher participants than are current agricultural programs.

Longtime agricultural policy expert Willard Cochrane has proposed that each "family farm" be awarded an annual payment of $20,000 per farm, the current annual authorization of $20 billion being adequate to provide such payments to two million farmers and ranchers.[9] Cochrane's proposal might be amended to provide for a $20,000 tax credit to go to each family farm that is demonstrating progress toward sustainability. A farmer with no net farm income would receive a $20,000 annual payment from the government to compensate them for conserving natural resources, protecting the natural environment, and contributing to the economic sustainability of their community.

Farmers who are approved for the tax credit would also have an alternative farm tax rate—possibly 50 percent of net farm income compared with the 15 to 20 percent typically paid by farmers today. Thus, as net farm income increases, the advantage of the tax credit would diminish as the higher tax rate claims a larger total amount. At a net farm income of $40,000, for example, the taxes owed (50 percent of $40,000) would completely offset the $20,000 tax credit, and thus the farmer would not pay in or receive anything from the government. At some higher level of income, probably between $60,000 and $80,000, it would be advantageous for the farmer to forgo the special farm tax credit and pay taxes as any other business. At this point, however, the sustainable farming or ranching operation would be sufficiently profitable to ensure its sustainability without any further government support.

Farmers and ranchers would be free to own and operate as many acres and to produce as much as they choose, but the tax credit would be limited to $20,000 for each full-time independent farmer. For part-time farmers, nonfarm income could be added to farm income for tax purposes, reducing the advantage of the farm tax credit in proportion to their reliance on nonfarm income.

No one would dictate who should produce how much of what products. Those decisions would be made by farmers and ranchers, not by the government and not by multinational corporations. Farmers and ranchers who chose not to participate in the long-run food security program would not be required to have a sustainability transition plan but would not be allowed to exploit their land or to degrade the natural environment. Such farms would be classified as industry, rather than agriculture, and would be subject to the same environmental regulations as any other producer of industrial commodities. No one has the right to exploit either the land or the people for short-run economic gain.

Such a program could be administered as a Farm Tax Program rather than a farm bill. Farm policy would then be transformed into food security policy, allowing it to be taken out of USDA, where the agricultural establishment has the power to block the redirection of benefits to serve the public good. The Farm Tax Program would provide farmers and ranchers with many of the employment security benefits available to other public workers, such as minimum wages, unemployment benefits, and workers' compensation. They would have the assurance of the tax credit to help them cope with years of crop failures, depressed prices, ill health, or other economic setbacks during their transition to sustainable farming. Over time, farmers and ranchers would be required to show progress toward sustainability to remain eligible for the tax credit. If, after some specified number of years, they fail to achieve economic sustainability, they could be helped to find employment elsewhere, freeing up their farms for a beginning farmer, who would then be eligible for the Farm Tax Program.

The principles guiding U.S. agricultural trade policies for a sustainable global society should be simple and straightforward. A truly effective World Trade Organization would empower every nation with both the right and the responsibility of protecting its natural resources and its people from economic exploitation. People within nations should be allowed to decide the conditions under which they choose to trade and choose not to trade, without threats or coercion.

The natural ecosystem is global, not national, and thus, all nations have a responsibility to ensure that the global environment is protected for the benefit of all people of the world. Increasingly, all nations share in a global culture, but global culture need not, and should not, erase all cultural differences among people. No nation has the right to impose the values of their culture upon other cultures of the world. The economy is increasingly global in nature, and there is much to be gained from trade among nations. But the removal of all national economic boundaries would inevitably lead to economic exploitation of the weak and the poor by the strong and the wealthy and to economic exploitation of the natural environment. The only truly free trade is trade among people who are truly free *not* to trade. U.S. trade policy should respect this right to ensure world trade that is truly beneficial to all.

Other government programs, including publicly funded research and education, could be redirected to support sustainable farming—to provide true public benefits rather than support private-public partnerships. State and federal programs could also be targeted to developing the physical and informational infrastructure needed to support local niche markets needed for sustainable-sized farms—connecting local consumers with local farmers. Federal, state, and local governments could be required to purchase agricultural products for schools, prisons, and other public institutions from local sustainable farmers, thus enhancing their chances for success. Government stocks of grains and other storable commodities could be held in farmer-owned facilities to keep them in the local community and enhance farm income as well. The justification for local purchases would be to provide maximum total public benefits rather than minimizing the cost of one public program at the expense of another.

Skeptics may question whether we can afford to abandon public support of large-scale corporate agriculture in favor of sustainability. Surely food costs will go higher, they claim, and consumers will revolt. However, such contentions are not supported by facts. Americans spend little more than 10 percent of their disposable income for food—a dime of each dollar. Equally important, only two cents of each dime they spend goes to the farmer who produces the food—eight cents goes for packaging, transportation, advertising, and other marketing services. Even if farmers required 10 percent more to produce sustainably, for example, food prices would only need to be 2 percent higher. Americans can afford a sustainable agriculture. But even more importantly, under corporately controlled agriculture, food prices in the future would likely be far higher than with sustainable family-farm agriculture.

As government programs targeted to long-run food security are developed and implemented, the productivity and economic viability of independent family farms will rise and the costs of government farm programs will fall. As ecologically sound and socially responsible farms become more productive and profitable, without government assistance, a sustainable agriculture will have permanently displaced the unsustainable industrial system that was based on exploitation of people and of nature. As industrial agriculture runs out of resources, places, and people to exploit, it will be surpassed in productivity and profitability by new sustainable systems of farming. Over the long run, a sustainable agriculture will feed more people better at a lower cost. By redirecting farm policies to focus on the public benefits of agriculture, Americans will have ensured the sustainability of their agriculture and their long-run food security.

|| Presented at the Symposium on Sustainable Animal Agriculture, Annual Meeting of the American Society of Animal Science, Phoenix, Arizona, June 23, 2003.

19

The New American Food System

The twentieth century was the American Century—as is commonly conceded by historians. During the twentieth century, the United States replaced Great Britain as the dominant global economic power and America's corporate version of capitalism replaced both socialism and classical capitalism as the world's dominant economic model. The United States came from behind to beat the Soviet Union to the moon and take leadership in space. The United States came from behind to pull ahead of Japan in electronics and communications technologies. And the United States replaced the whole of Europe as the single dominant global military power.

The American Century was a time during which economics gained precedence over all else—including politics, society, and culture. America struggled economically along with the rest of the world during much of the first half of the century. But America built the foundation for its modern industrial economy during World War II, used its postwar economy to help rebuild Europe and Japan, and thereafter never looked back. The United States' desire for maximum economic growth provided the motive for its unrestrained corporatist economy, which later became the model for developing economies around the globe. Research and development supported by economic growth allowed America to take world leadership in space and electronics. And economic growth made possible the most powerful and dominant military force ever assembled in the history of humanity.

As we enter a new century, however, public concerns are growing regarding the sustainability of the country's economic growth. Growing evidence of air and water pollution during the 1960s raised questions concerning the inherent negative environmental impacts of the industrial paradigm of economic development. The energy crisis of the 1970s raised concerns about the extractive nature of the "free market" economy, and its inherent reliance on limited supplies of nonrenewable resources. The economics of greed, which characterized America in the 1980s, raised new concerns about a growing economic gap between the "haves and have-nots." And when the economic bubble of the 1990s burst at the turn of the century, many more people began to question whether America's economic growth was sustainable.

The environment has been the focus of primary concern for sustainability, but there are growing questions of social and cultural sustainability as well. Our relentless pursuit of economic prosperity is separating people within families, communities, and society as a whole and is destroying the social fabric of our country. The health of any society is reflected in the quality of relationships among its people—within families, communities, and society in general. During the latter half of the twentieth century, as American society has become increasingly disconnected, our relationships have become increasingly unhealthy and dysfunctional. In our quest for global economic supremacy, the United States has become a splintered nation of disconnected people.

We Americans have come to deal with each other only indirectly—through markets, through agents, or through lawyers and courts. Our relationships are defined by transactions, contracts, and laws rather than by common interests, commitments, and trust. In the marketplace, we are committed to competition, not cooperation. We take adversarial positions in the courts in our search for truth. Our personal disagreements lead to arguments and threats, and we settle our international disagreements through coercion and "small wars." Relationships based on believing, trusting, caring, and sharing are labeled as naive or idealistic. We seem to be a nation that has lost any sense of personal connectedness.

In his book *Bowling Alone*, Harvard political scientist Robert Putnam provides measure after measure verifying that Americans have became increasingly disconnected during the last half of the twentieth century.[1] Fewer Americans voted in elections, belonged to organizations, participated in social activities, visited each other's homes, or did other things necessary to build personal relationships. Most such measures of social connectedness have dropped by 30 to 50 percent since the late 1950s. Putnam says that we remain interested and critical spectators of the public scene, but we don't play. We remain affiliated with various civic associations, but we don't show up. We attend public meetings less often, and when we do, we are disappointed to find that few of our neighbors have joined us. We are less generous with our time and money, we are less likely to give strangers the benefit of a doubt, and they return the "favor." Between the 1970s and 1990s, the numbers of lawyers per person in the United States more than doubled. We spent 40 percent more for police and security guards and 150 percent more for lawyers and judges than would have been expected based on growth in population and the economy between the 1970s and 1990s. As Americans have become disconnected, we apparently have become a more contentious, less civil society.

The term "social illness" is more than a convenient analogy in this case. Putnam points out that the rate of mental depression among the past two generations in America has increased roughly tenfold—these being the generations most socially disconnected. It might be tempting to attribute this rise to a greater willingness to acknowledge depression; however, between 1950 and 1995, the rate of suicide among American adolescents more than quadrupled and among young adults nearly tripled. Suicide and clinical depression, fortunately, are not all that common among the general population. However, incidents of "malaise"—headaches, indigestion, and sleeplessness—are far more common and show patterns similar to more serious mental illnesses. Surveys between the late 1970s and late 1990s indicate that those of each new generation suffer from higher levels of "malaise" and on average are "less happy" than are those of previous generations. As each generation has become increasingly disconnected, the nation as a whole has become increasingly mentally ill and physically miserable.

It's no coincidence that people have become increasingly disconnected from each other, as well as from the earth, during the last half of the twentieth century—the last stages of industrialization. Disconnectedness is an unintended but inescapable consequence of the industrial approach to economic development. The principles of industrialization are the same for automobile manufacturers, large-scale vegetable processors, retail superstores, and confinement animal feeding operations. The gains in efficiency from industrialization are achieved by carrying out specialized functions by standardized means under centralized management. In such systems, relationships must be impartial and thus impersonal, and we must compete rather than cooperate if maximum economic efficiencies are to be achieved. Thus our growing social disconnectedness, resulting from loss of personal relationships, is not a coincidence but a direct consequence of American industrialization.

Nowhere in the United States is this social disconnectedness more evident than in our systems of food and farming. Most consumers, particularly younger consumers, have no sense of where their food actually comes from or who produces it. Even those who understand that farmers grow crops and livestock, others process and package these products, and still others deliver food to grocery stores and restaurants, have little sense of what's actually involved in these processes. We shouldn't be surprised that consumers have no real understanding of food, because they have no sense of connectedness with the land or with the farmers who tend the soil.

Before industrialization, when the United States was an

agrarian nation, people produced their own food, bartered for food, or bought food from someone who had produced it. The relationship between consumer and producer was direct and personal. As the economy became more specialized, merchants such as butchers, bakers, and brewers bought from producers and sold to consumers, and the farmer-consumer connection became one step removed. Later, grocery store owners bought from the butchers, bakers, and brewers, and then consumers were at least two steps removed from the farm.

As the food system moved beyond the early stages of industrialization, control of the system began to be consolidated in the hands of a few large food corporations. New industrial technologies and organizational models required still larger capital investments. First, independent entrepreneurs were displaced by family corporations, but eventually, few families could accumulate enough capital to compete. As market power and political power replaced economic efficiency as the primary motivation for consolidation of control, only the giant publicly held corporations were able to compete.

In farming, independent family farms were replaced by family corporations, which are now being replaced by corporately controlled contract production—factory farming. In food retailing, the "mom-and-pop" corner grocery stores were displaced by regional and national chains of large supermarkets, which now are also being displaced, by global chains of even larger retail "supercenters." Independently operated restaurants and delis were displaced by franchised restaurants and fast food joints. Independent food processors and wholesalers were displaced by giant food processing and distribution firms, which since have been absorbed into five or six even larger "global food chain clusters."[2] Most people today don't have a clue as to where in the world their food comes from, how it was produced, or who produced it.

But does it really matter if people don't understand where their food comes from, or if they think it is manufactured rather than grown? People don't understand where their automobiles, clothes, houses, movies, or much of anything else comes from, and no one seems to be complaining about the lack of knowledge of such things. However, all disconnections among people matter, even if no one complains. The seeds of dissension are sown in the gaps of understanding and appreciation that exist among people. Conflict, frustration, depression, malaise, and many other miseries of life are but symptoms of our lack of understanding and appreciation for each other. People may not have associated the symptoms with the cause, but the cause still matters. And it matters even more that we consumers understand our connections with farmers.

Many farmers feel a great sense of frustration that people don't understand how life in general is connected to life in the soil and the life of people who till the soil. They feel they are forced to destroy the natural productivity of the soil, to degrade the natural environment, and to destroy the social fabric of their communities because the only things consumers are concerned about are price, convenience, and cosmetic appearance. Many farmers feel that they are forced to value the economic bottom line above virtually all else, above their neighbors and communities, and sometimes even above their families, because they believe the only thing consumers care about is "cheap food." Farmers want to be good neighbors and good stewards of the land, but the competitive pressures of a consumer-driven market economy won't let them. Instead, the land, the quality of rural life, and ultimately the ability of the earth to support human life will be destroyed because of the disconnectedness of Americans from the land and from the people who farm it.

Unfortunately, the only link between farmers and food consumers is a disconnected, dysfunctional, and unsustainable food system. As a prime example, Eric Schlosser, in his recent best seller, *Fast Food Nation*, attempts to assess the social cost of our "love affair" with fast foods.[3] Food eaten away from home now claims a share approaching half of all food purchases in America. And "fast food" places such as McDonald's, Kentucky Fried Chicken, Taco Bell, and Pizza Hut account for nearly half of all food consumed away from home. Schlosser writes that fast food has triggered the homogenization of our society, hastened the "malling" of our landscape, widening of the chasm between rich and poor, fueled an epidemic of obesity, and propelled the juggernaut of American cultural imperialism abroad. He documents how fast foods have lured us into choosing diets deficient in nearly everything except calories, supporting practices deceptive in every aspect, from advertising to flavoring, and promoting systems that degrade nearly everyone and everything involved in the process.

The fast food industry has lured low-income consumers, along with the affluent, into paying ridiculously high prices for low-quality meats, potatoes, vegetable oil, and sugar. However, the high dollar-and-cent costs are just the tip of the iceberg. The true costs of fast food must include the costs of poor health, lost dignity in work, degraded landscapes, and ethical and moral decay in business matters, including international trade and investment.

With the rapid consolidation now taking place among food supermarket chains, the fast food story undoubtedly has relevance for the whole of food retailing. The independent food processors, distributors, and retailers today are under the same economic pressures as independent family farmers. They are fighting for their economic survival. They can't afford to be too concerned about the well-being of their employees, their suppliers, or their customers; they have to look out for themselves. If their labor costs are too high because of generous salaries and benefits, they can't compete. If they pay too much to farmers or other suppliers of raw materials, their profit margins will disappear. If they don't take advantage of the natural human frailties of their customers, their competitors will. If a store or processing plant isn't profitable in one community, they feel compelled to move to another, regardless of the impact on the community.

The independent food marketer, like the family farmer, is in a struggle for economic survival.

Thankfully, there are signs of change on the horizon. A new American culture is emerging to challenge the current industrial culture of economic materialism. In their book *The Cultural Creatives*, Paul Ray and Sherry Anderson provide compelling evidence that some fifty million Americans are now leading the way in creating a new American culture.[4] The authors identify three distinct groups of American adults—moderns, traditionalists, and cultural creatives—based on some 100,000 responses to surveys concerning basic values and lifestyles and numerous focus groups and personal interviews.

The "moderns," the dominant group, tend to define the public perception of American society. Their values are reflected in a preoccupation, if not obsession, with material success—making money, getting ahead, looking good, and living the affluent lifestyle. The moderns care about family and community and have some concern for the natural environment, but they care far more about their individual success.

The "traditionalists" make up about a quarter of the adult population. The authors describe the traditionalists as wanting the world to be "like it used to be but never was." They care about community and family values, but their primary focus is on restoring culture to some idealized vision of earlier times.

They profess a commitment to community and stewardship, but tend to limit their concerns to their circle of friends and their property and express less concern for the natural environment than do either of the other two groups.

The "cultural creatives" are distinguished from the other two by their strong belief in the value of personal relationships, within families, communities, and society as a whole, and by their concern for the integrity and long-run sustainability of the natural environment. They are found in association with various social movements, including social justice, environmental protection, civil rights, gender rights, and sustainable development. They are less materialistic than are either of the other groups, and they tend to be more spiritual, in the sense of a personal connectedness with something beyond self.

The cultural creatives group made up about 27 percent of those surveyed but was growing rapidly and could easily make up one-third of American adults today. The dominant group, the moderns, made up about half of those surveyed, but only about half of this group was firmly committed to the materialistic principles of individual economic self-interest. About a quarter of the moderns, one-eighth of the total, were too busy trying to get ahead or to make ends meet to think about what they believe. The remainder of the moderns actually felt alienated by modern society; it wasn't working for them. They were going along because they didn't see any viable alternative. The traditionalists made up almost a quarter of those surveyed. They share some of the values of material self-interest with the moderns but also share some of the spiritual values of the cultural creatives.

The values and lifestyles of the cultural creatives are completely consistent with the principles of sustainable development and sustainable agriculture. They believe that quality of life results from equitably meeting the needs of the present while leaving equal or better opportunities for the future. The sustainable development movement arose from a growing realization that economic development alone does not increase overall quality of life but instead often leads to its degradation. They believe that balance and harmony among the ecological, economic, and social dimensions of life define what it means not only to live sustainability but also to live a life of quality.

This newly emerging American culture is based on the realization that standard of living is only one dimension of the quality of life. Those creating the new culture are pursuing a more enlightened concept of self-interest. They recognize that people have broader interpersonal self-interests and higher spiritual self-interests in addition to narrow individual self-interests. The new American culture reflects a realization that true self-interest depends on balance and harmony among these three layers of self.

This enlightened self-interest is not some New Age radical concept. In the early 1800s, Alex de Tocqueville wrote in his classic book, *Democracy in America*, Americans believed strongly "that men ought to sacrifice themselves for their fellow-creatures . . . that such sacrifices are as necessary to him who imposes them upon himself as to him for whose sake they are made."[5] Tocqueville called this belief "self-interests rightly understood." It recognizes the fact that people benefit from fulfilling their proper role in the larger society in ways that could never be linked directly to one's narrowly defined, individual self-interest. He believed that such a culture was necessary to restrain our greed and to sustain the American democracy. Thankfully, America appears to be returning, although slowly, to those cultural roots.

The ranks of the cultural creatives include thousands of American farmers. These culture-creating farmers may call themselves organic, biodynamic, holistic, natural, ecological, practical, or just plain family farmers. However, these new ways of farming all fit under the conceptual umbrella of sustainable agriculture. The sustainable agriculture movement is a small but critical part of a much larger movement promoting sustainable development. A recent publication of the USDA Sustainable Agriculture Research and Education program highlights fifty such farmers from across the United States.[6] Thousands more, each with a unique and different story, share a common vision for a brighter, more sustainable future for agriculture. These new farmers experience many frustrations and hardships along with the joys of success. Creating a new culture isn't easy—on farms or anywhere else—but more and more of these new farmers are finding ways to succeed.

In general, sustainable farmers succeed by focusing on the weaknesses of industrial systems of food and farming. They realize economic gains from appropriate levels of specialization, standardization, and consolidation but without sacrificing the social, ecological, and economic benefits of positive relationships among diverse elements of unique, interdependent systems. They don't compete with industrial agriculture; they do something different. They focus on doing the things that industrial systems are inherently incapable of doing well.

They are rediscovering the fundamental roots of agriculture; they are reconnecting to the land and to each other, and in the process are redefining farming. There are no blueprints or recipes for sustainable farming; however, some general underlying characteristics of successful sustainable farming operations are beginning to emerge from the diverse experiences of these new farmers.[7]

First, these farmers see themselves as stewards of the earth. They have a deep sense of personal connectedness to their land. Their farming operations tend to be more diversified than conventional farms because nature is diverse and they create farms that respect the diversity of the land. They find that when they work *with* nature, nature is productive, and their farms are made economically viable as well as more ecologically sound by reconnecting with the land.

Second, these new farmers build relationships. Most of these new farmers establish personal connections with their customers. They connect with customers who care where their food comes from and how it is produced, and they receive premium prices by producing foods that are valued by their particular customers. Their farms are made profitable as well ecologically and socially responsible by reconnecting with their customers.

These new farmers also reconnect with each other—to buy equipment, process and market their products, and do other things that they can't do as well alone. Their relationships go beyond economic agreements; they establish a sense of personal connectedness. They have learned that such personal relationships are necessary to sustain their economic, ecological, and social quality of life.

Finally, these new farmers are quality-of-life farmers, and their quality of life depends on the quality of their relationships. To them, the farm is a good place to live, a good place to maintain healthy connections within families, and a good way to reconnect with people within communities. Their connections are not just social but also spiritual, both with other people and with the earth, and they reap the benefits in terms of a strong sense of purpose and meaning in their lives. They farm to make a living, but more importantly, they are farming for a higher quality of life. These new American farmers are agriculture's cultural creatives.

Independent food processors, distributors, and marketers also are beginning to realize they face the same kinds of challenges and have the same kinds of opportunities as independent family farmers. Independent food marketers cannot expect to compete with the giant global-food-chain clusters and are too small to form their own strategic alliances to compete in the global arena. If there is to be a future for independent food processors, distributors, or marketers, it will be outside the global food chain. They must connect with sustainable farmers to create a new sustainable American food system.

Independent retailers and restaurants must learn to market in the niches—meeting the needs of consumers that are not being met by the industrial, mass-production, mass-distribution food system of today. Many consumers today don't trust the current food system. They are concerned about food safety and nutrition and are dissatisfied with the taste and flavor of many industrial food products. They will pay premium prices for wholesome, nutritious food that really tastes good. Many will pay premium prices for crops that are grown organically or for meat from animals raised under humane conditions, without chemicals, without hormones or antibiotics. The mass-production, mass-distribution food system cannot meet the unique needs of the cultural creatives—at least not as effectively as can the smaller, individually owned, personally managed food business. A new cultural-creative food system, separate from the industrial food system, must emerge to meet the needs of a growing cultural-creative food market.

The skeptics claim that such markets are inherently small and limited in importance. In reality, all consumer markets are niche markets, because all people have somewhat different tastes and preferences. The mass-market merchandisers attempt to target the middle of the distribution of consumer preferences, where individual tastes and preferences are similar but never identical. As more consumers become increasingly dissatisfied or disenchanted with industrial mass-produced foods, the opportunities for accommodating those diverse individual tastes and preferences will continue to grow.

The Hartman Report—a professionally designed and conducted survey of representative U.S. households—explored how consumers' food purchases are actually affected by their environmental attitudes.[8] The report identified two groups, the "true naturals" and "new green mainstream," which make up about 28 percent of the population, as prime markets for sustainably produced foods. These groups are willing to act on their preferences by paying premium prices for sustainably produced foods. The groups are very similar in attitudes and magnitude to Ray and Anderson's cultural creatives. Armed with the ecological, social, and economic facts of today's food system, and an opportunity to choose a sustainable alternative, an even larger group of consumers almost certainly would be willing to pay the full economic costs of a truly sustainable food system.

The current challenges for independent processors, distributors, and retailers are real, but their opportunities are unlimited. Most consumers today are simply unaware that a handful of multinational corporations are quickly gaining control of the global food supply. Millions of consumers would be willing to pay the cost of an independently owned and operated food system, if they realized the consequences of not doing so.

The new American food system will be dramatically different from today's industrial food system. Quite likely, it will be a network of local, interdependent community food systems rather than part of some corporately controlled global-food-chain cluster. The key to success in the new food system will be relationship marketing. Relationship markets are built on personal connectedness, and such connections are far easier to establish and maintain where farmers, processors, retailers, and customers all live in geographic proximity. "Local" is becoming the "new organic," as industrial organics make up an increasing share of organic foods in mainstream restaurants and supermarkets.

Restaurants seem to be leading the trend toward buying local. The Chefs Collaborative, made up of chefs from upscale restaurants throughout the country, is the most prominent example. One of their fundamental organizational principles is "Sound food choices emphasizing locally grown, seasonally fresh, and whole or minimally processed ingredients."[9] Their other principles are very much in harmony with the development and support of an ecologically sound and socially responsible food system. Independent restaurants everywhere, across all price ranges, seem to understand that their best defense against the national franchises is to advertise their reliance on local farmers who provide them with really fresh, high-quality foods.

Another organization giving voice to the growing preference for a network of community-based food systems is Slow Food. Slow Food is a worldwide movement committed to promoting the diversity of local and regional quality food produced and marketed in a way that guarantees farmers a fair price and protects the environment and the natural landscape.[10] Those in the movement have a clear understanding of the industrial food system and they realize that a return to local and regional food systems will be necessary for ecological and social sustainability. In his book, *The Pleasures of Slow Food*, Corby Kummer points out that Slow Food is not an elitist gourmet movement but instead encourages "good, honest food at reasonable prices" and its appreciation and enjoyment to the fullest by all.[11]

While these movements still may be small, they are helping to create a new food culture for the future. The cultural creatives within society are just beginning to realize that they can reflect their values and pursue their preferred lifestyles through their food choices. As the availability and awareness of alternatives to industrial mass-produced foods become more common, demand for something fundamentally different and better will continue to grow. The cultural creatives didn't exist forty years ago and perhaps accounted for 5 to 10 percent of Americans a decade ago; today they may account for one-third or more of the total population, and they are still growing. Current sales of organic, natural, socially responsible products represent but a small fraction of the current potential market represented by this large and growing segment of American society.

The cultural "moderns" of American society are not yet ready to participate in creating a new American food system. Their values and lifestyles must change before they will be either willing or able to become part of the new food culture. Unfortunately, some advocates of a more sustainable food system tend to spend far too much time and energy worrying about how they can change the moderns rather than how they can meet the needs of the cultural creatives. The new American food system will be developed by a coalition of farmers, marketers, and consumers who already share the values and aspire to the lifestyle of the cultural creatives.

Disenchanted moderns and traditionalists will join the new culture only when it is demonstrated to be a better way to live—a higher quality of life. The primary obstacle in creating a new American food system today is not the values and lifestyles of the moderns or traditionalists. Instead, the larger obstacles are a lack of awareness among food consumers of a sustainable alternative and a lack of a sufficient number of sustainable farmers who are willing and able to produce for the market that already exists.

Sustainability is about moving forward to something better, not going back. The goal in creating local food systems is not to return communities to self-sufficiency in food production, any more than the goal of sustainable agriculture is to make farms self-sufficient. There are real and significant benefits to be gained from relationships with others, both socially and economically; relationships of choice can be mutually beneficial. The goal is to learn to work and live in harmony with nature—including human nature—in order to build positive relationships among people and between people and the earth.

As a means of achieving this harmony, the new American food system will encourage and support production of foods uniquely suited to specific ecological and cultural niches. It will also encourage and support local consumption of local foods, in the belief that eating foods produced in the places where we live, by people we know, is an act of integrity and value. The fundamental purpose of local community-based food systems is to reconnect us to the earth and to each other. However, this connection does not imply that consumers should consume only food produced locally or that farmers should sell all of their products locally. Community food systems only require a "preference for the local" as a means of reconnecting with our neighbors, and thus enhancing our quality of life.

Many may question whether these local community-based food systems can possibly replace the corporate industrial food system of today. Actually, networks of interdependent community-based systems of the future might serve the total food market more easily, efficiently, and effectively than can a giant, hierarchically managed, corporately controlled, and centrally planned global food chain. Local community systems could be quite easily linked through formal and informal arrangements so that surpluses could be shared, first within regions, then within nations, and finally among regions and nations of the world. Each community food system might operate something like the merchants' guilds of earlier times. However, unlike the merchants' guilds, community food systems would include consumers as well as producers, and would recognize the necessity for sharing, among communities and across regions, in achieving a desirable quality of life.

Some food merchants might choose to form organizations to vouch for the integrity of its members, although each member offers unique, location-specific foods, as seen currently in global "fair trade" of coffee, bananas, and dozens of other foods. The result might be a global food network, but one that reflects a strong preference for things local, and thus, things that keep us most connected with the earth and with each other. Such a food system would reflect our pursuit of a more enlightened self-interest and a more desirable quality of life.

But is it realistic to expect, or even to hope for, such a radically different food system? In his best-selling book, *The Tipping Point*, Malcolm Gladwell relates dramatic changes within society to the spread of a disease epidemic.[12] He contends that the seeds of radical change are always present within society, but they only spread and eventually explode into an epidemic of change under specific circumstances. He identifies three rules of epidemics, or three preconditions for explosive change. First, for a new idea to spread, the idea needs to "infect" people who are effective in infecting others—people he calls "connectors." Connectors are those who have contacts and influence with many other people. Next, for an idea to grow it needs to catch on and hold on in the minds of those who have been infected—it has to be "sticky." Finally, in order for an idea to break out into an epidemic of change, it must have an accommodating environment or social context—society must be ready for change.

American society is ripe for an epidemic of change from the old industrial to a new sustainable society. And this epidemic will bring with it a new American food system. Advocates for environmental protection and social justice are no longer on the fringes of society. They are prominent among educators, writers, religious leaders, actors, and even some leaders in business and politics. The Internet provides an unprecedented tool that allows even ordinary people to connect with thousands of others—quickly, frequently, and inexpensively—and thus multiplies the number of social connectors. And if we can break the grip of corporate influence on politics and business, advocates of a truly sustainable society will be at least as prominent in politics and business as in everyday life.

Until recently, the messages of environmental protection and social equity had been interpreted as messages of sacrifice. We of the present must sacrifice for the benefit of those of the future; those who *have* must sacrifice for the benefit of those who *have not*. But Americans are awakening to the reality that our quality of life has been diminished by our exploitation of the environment and of each other in the pursuit of our narrow, individual self-interests. Americans are beginning to realize that taking care of the earth and taking care of each other are not sacrifices but instead enhance our quality of life. The pursuit of quality of life instead of standard of living is a "sticky" message that will cling to the minds of all who understand it.

Finally, the current social context is ripe for the outbreak of an epidemic of change. Most people realize that the industrial era is over and a postindustrial era is upon us; we don't know what to call it yet, but we know it will be different. The economic bubble of the Reagan-to-Clinton era has burst, the world is still lingering on the verge of recession, and no one knows how far it is to the bottom if we eventually fall over the brink, or even if the economy could ever recover. The world has been dragged into a global "war on terrorism"—a war that apparently will be punctuated by periodic "small wars" and admittedly has no foreseeable end. American society will only tolerate this continuing uncertainty and vulnerability for so long, and then it will demand fundamental change. Americans will reject the current modern values and lifestyles and will embrace the creation of a new American culture.

According to Ray and Anderson, more than half of today's modern materialists are disenchanted—either alienated, but without an alternative, or striving, with little hope for success. Both of these groups may quickly become disenchanted with their uncomfortable allegiance with the core moderns and join with the cultural creatives to explore new directions in pursuit of a better quality of life. The traditionalists may eventually realize that the principles they now seek through religion are found in commonsense principles of sustainability—in applying the Golden Rule to all people, both within and across generations. American society shows all the signs of a society ready for an epidemic of radical, fundamental change.

Now is the time to create a new American food system—a network of community food systems linking independent, local farmers with independent, local food processors and retailers, to provide food for customers willing to pay for quality and integrity. It's time to create a food *values* chain linked by the principles of ecological integrity, economic viability, and social responsibility. This task will take time and effort to complete, but now is the time to begin.

The new food system will reconnect people with the earth and with each other, and thus will contribute to a more enlightened concept of quality of life. In creating this new and better food system, a sustainable food system, we will be leading the way to a brighter, more sustainable future for America and for the rest of the world.

॥ Presented at the Ohio Ecological Food and Farming Association Twenty-third Annual Conference, Johnstown, Ohio, March 8–9, 2003.

20

American Agriculture After Fossil Energy

The world is running out of *cheap* fossil energy. Some dismiss the current energy crunch as nothing more than a short-run phenomenon, arguing that we have used only a small fraction of the earth's total fossil energy reserves. While this argument contains an element of truth, it masks far more than it reveals. The industrial era has been fueled by *cheap* energy, first by wood from abundant forests and then by fossil energy from easily accessible sources. But the days of old-growth forests, oil gushers, and surface veins of coal are gone. Most of the remaining reserves of oil and natural gas are buried far below the earth's surface or deep beneath the ocean floor. The remaining reserves of coal likewise will be more costly to mine and also more costly to burn without polluting the air and degrading the environment. There are no more sources of cheap fossil energy. The industrial era that has characterized modern society for the past two hundred years is coming to an end.

The concept of "peak oil" has gained wider public attention over the past few years as prices for petroleum have climbed.[1] Petroleum geologists observed several decades ago that the peak in production from a given oil field typically occurs when approximately half of the recoverable oil in the field has been extracted. After the peak, production continues but only at a diminished rate. Historically, the time lag between discovery and peak production has averaged about 30 to 40 years. It takes time to get started drilling and time to drill a sufficient number of wells to reach peak production. Beyond the peak, the old wells yield less oil, and as residual reserves decline, new wells typically are deeper, more costly, and less productive.

Oil discoveries peaked in the United States in Oklahoma and Texas in the late 1930s and early 1940s. In 1971, U.S. petroleum production peaked and has been declining ever since.[2] The vast new oil fields in Alaska caused but a temporary "blip" in a persistent downtrend in production. In spite of rhetoric to the contrary, the United States has been powerless to reduce its growing dependence on foreign oil by increasing domestic production. The peak in global oil discoveries occurred in 1962, which would indicate a peak in global production sometime in the early 2000s. Experts disagree about when the peak will actually occur, with estimates ranging from as late as 2025 to as early as 2005. Global production has been essentially flat since 2005, in spite of record oil prices, so the peak may have already been passed. Even the major oil companies, such as BP, Exxon Mobil, and Chevron Corporation, have begun to focus their attention on energy alternatives for the future.

The experts generally agree that we have not come close to depleting the earth's petroleum reserves. However, about half the total reserves are considered to be nonrecoverable, using any known technology. Still, we have only used about half of the estimated *recoverable* oil reserves. The problem is that recovery costs will increase and annual production will decline as the remaining recoverable reserves are diminished. Even if new technology is found to recover more of total reserves, recovery will still likely be slower and more costly.

The logical alternative sources of fossil energy also are all more costly than petroleum produced from existing oil fields. The inevitability of increasing costs can be seen most clearly in the relative amounts of existing energy required to extract new energy from various alternative sources. Energy is required to drill, mine, extract, crush, distill, refine, and carry out the other processes necessary to turn energy reserves into usable energy. Regardless of changes in dollar-and-cent costs, alternative energy sources that require more existing energy to create new usable energy will be more costly.

Oil produced in the United States during the 1940s yielded more than 100 kilocalories of energy for each kilocalorie of energy used in extraction, a net energy ratio of over 100 to 1.[3] By the 1970s, with deeper, less productive wells, the ratio had dropped to 23 to 1. For today's production from oil discoveries made during the 1970s, the ratio has dropped to only 8 to 1 kilocalories of new energy for each kilocalorie of existing energy used to produce it. Falling net energy ratios and rising energy costs have made alternative sources of petroleum competitive with current production. For example, oil from tar sands in Alberta, Canada, are attracting increased investments, in spite of net energy ratios of less than 8 to 1. Liquefied coal, which was used by Hitler to fuel the Nazi army during World War II, also has a net energy ratio of about 8 to 1. Oil shale, although abundant in supply, presents an even more formidable challenge from a net energy perspective.

In addition, all of the most competitive alternatives to oil raise far more serious environment risks than do oil production and refining. For example, replacing existing petroleum usage with oil from coal would add large amounts of greenhouse gases to the atmosphere, especially carbon dioxide, at a time of growing concern about global warming. Replacing current global usage of crude oil with tar sands would require a waste pond equal to half the size of Lake Ontario to accommodate the toxic liquid waste.[4] Further exploration and drilling offshore and in national parks and wildlife reserves threaten ecological destruction—for nothing more than another small blip in the downtrend in energy production—at a time when long-run sustainable energy production may well depend upon healthy biological systems.

All the other alternative sources of fossil energy face futures very similar in nature to petroleum. Natural gas supplies may be the next to peak after oil and the timing of the peak in coal production will depend to a great extent on whether environmental issues are resolved, either through new technology or by relaxing environmental standards. If coal is used to replace the shortfalls in oil and natural gas, within fifty years the energy obtained by extracting fuel from the coal might well be less than the energy required to mine the coal. The world isn't running out of fossil energy, at least not yet, but it *is* running out of cheap fossil energy.

New technologies may be found to extract more energy quicker, turning energy peaks into plateaus and recovering energy reserves now considered nonrecoverable. But such technologies would simply sharpen the ultimate drops in fossil energy production, as total reserves would be more quickly depleted. All economically recoverable *nonrenewable* energy resources eventually will be depleted. It's just a matter of time. In addition, global population is projected to increase by half again within the next fifty years, and booming industrial economies in the two most densely populated countries of the world, China and India, promise to increase global energy demand far faster than growth in global population. Fossil energy production will almost certainly fall far short of meeting this growing global demand, regardless of new technologies.

All of the alternative energy sources—nuclear, wind, water, photovoltaic—will be more costly than today's fossil energy, both in terms of net energy produced and dollar-and-cent costs. Gasoline prices may vary with changes in economic and political conditions around the world, but the underlying trend will be rising energy costs into the indefinite future. Cheap and abundant energy has shaped the past two hundred years of human society—the industrial era. The next two hundred years may well be shaped by the scarcity and high cost of energy.

A new "oil boom" in agriculture has been sparked by rising petroleum costs and fueled by prospects for even higher energy costs in the future. Ethanol and biodiesel can be produced domestically from renewable sources, and these biofuels present fewer environmental threats than do most alternative sources of liquid energy. To many, biofuels seem to be an answer, if not *the* answer, to the United States' growing dependence on imported fossil energy. With the growing economic and human costs of U.S. military involvement in the Middle East, politicians have been quick to support any alternative to our continued reliance on Middle Eastern oil—the only major oil-producing region that has not yet peaked in production. Only the most naive believe that the cost of U.S. dependence on foreign oil is fully reflected in prices at the gas pumps. Biofuels have also been promoted as a source of employment and economic development for chronically depressed rural communities.

So ethanol plants have begun to spring up all across rural America, reminiscent of grain elevators along the new westward railroads of earlier times. In early 2006, the Renewable Fuels Association reported ninety-five ethanol plants already in operation, forty-six of which were farmer-owned, capable of producing four billion gallons of ethanol a year, with another thirty-one plants under construction.[5] This biofuels industry association estimated that ethanol had created 147,206 new jobs and added $14 billion to the U.S. gross domestic product. The USDA estimated that ethanol claimed 18 percent of the 2005 U.S. corn crop and seems certain to use even an even larger share of the 2006 crop. An agricultural oil boom clearly is under way.

But are biofuels really the answer, or even an answer, to the most important questions raised by rising energy costs? Admittedly, ethanol and biodiesel are alternative sources of *liquid* energy—the type of energy currently in shortest supply. If biofuels were simply promoted as such, there might be nothing misleading about their growing popularity or political support. However, biofuels can never significantly reduce U.S. reliance on imported oil. And perhaps most importantly, biofuels are not a sustainable source of either renewable energy or rural economic development. It's easy to understand why American farmers are willing to accept government subsidies for bioenergy production. Other businesses "root their way to the public trough," so why not farmers? But neither biofuels nor government subsidies offer realistic solutions to increasing U.S. foreign energy dependence or to the chronic economic crisis in rural America.

If all the solar energy collected by all the green plants in the United States could be magically converted into fossil energy, it would replace only about half of the fossil energy currently consumed annually in this country.[6] Agriculture accounts for only about one-third of all green plants, meaning that total solar energy captured by agriculture amounts to only about one-sixth of U.S. fossil energy use. In addition, only about one-fifth of solar energy captured by agriculture is harvested as food crops such as corn and soybeans. So the total energy in all food crops amounts to only one-thirtieth of total fossil energy use or about one-tenth of the total energy in U.S. petroleum use.[7] A recent Academy of Science report indicated that if the total U.S. corn and soybean crops were devoted to biofuels, ethanol could supply about 12 percent of current gasoline use and biodiesel about 6 percent of current diesel use.[8]

In addition, it takes fossil energy to produce agricultural crops and to transform those crops into biofuels. The net energy estimates for ethanol range from a net deficit, suggesting the fossil energy used in ethanol production exceeds the bioenergy produced, to a net energy surplus of about 50 percent. Biodiesel typically comes out somewhat better on the high side, with up to two kilocalories of energy produced per kilocalorie of fossil energy used. The net energy gains from producing ethanol and biodiesel from all food crops probably would amount to only about one-thirtieth, rather than one-tenth, of current petroleum use. Switchgrass, sugarcane, and other high-energy grasses have been suggested as sources of renewable energy. But even if the whole of agriculture were devoted to replacing fossil energy, biofuels would replace only about one-sixth of current petroleum use and one twentieth of total fossil energy.

Americans clearly are addicted to fossil energy. But Americans will give up air conditioning "mini-mansions" and their suvs before they give up eating. Thus, agriculture will not be devoted entirely to offsetting the decline in fossil energy production. Even if current net energy ratios are improved significantly, biofuels will never be a significant replacement for fossil energy. Biofuels are simply a means of converting some of the immobile energy used in agriculture, such as natural gas and electricity, into a more mobile, liquid form of energy. The current enthusiasm for biofuels, however, risks becoming a distraction from the more important task of developing renewable energy resources with far greater total potential, such as wind, water, and photovoltaic cells.

The highest priority for American agriculture should be on reducing the fossil energy dependence of food production. Our current food system, including food processing and distribution, claims about 17 percent of total U.S. fossil energy use, with about one-third of this total used at the farm level.[9] In fact, we use about ten kilocalories of fossil energy for every kilocalorie of food energy produced, not counting the energy use in final food preparation. Even at the farm level, American agriculture uses about three kilocalories of fossil energy for every kilocalorie of food energy produced. In a world of rising population and dwindling fossil energy, the first priority of agriculture should be producing more food with less fossil energy.

The fundamental purpose of agriculture is to collect solar energy and to transform it into forms that can be used to support human life. People simply cannot eat sunlight. Solar energy must be collected, converted, concentrated, and stored by green plants before it is useful to humans. Agriculture is quite capable of meeting the food needs of the global population of today, and possibly feeding twice or even three times as many people in the future, even while reducing its reliance on fossil energy. However, achieving this objective will require a fundamentally different kind of agriculture. The industrial agriculture of today is not sustainable.

The lack of sustainability of industrial agriculture is not a matter of personal opinion, it is a logical consequence of the most fundamental laws of science, the laws of thermodynamics. Sustainability ultimately depends on our use of energy, because anything that is useful in sustaining life on earth ultimately relies on energy. All material things that are of any use to us—our food, clothes, houses, automobiles—require energy to make and energy to use. Actually, all material things—such as food, gasoline, wood, plastic, and steel—are concentrated forms of energy. All useful human activities—working, managing, thinking, teaching—require human energy, which comes from the physical energy in the things people use. Physical scientists lump all such useful activities together and call them "work." Thus, all work requires energy.

In performing work, energy always changes in form—specifically, from more-concentrated forms to less-concentrated, more-dispersed forms. In fact, this natural tendency to disperse gives energy its ability to perform work. Energy is dispersed when matter is changed into energy, as when we eat food or burn gasoline. Energy also is dispersed when heat is used to produce electricity and electricity is used to produce light. However, regardless of the form of energy or the work it performs, the total energy embodied in matter and energy always remains unchanged. This is the law of energy conservation, as in Einstein's famous $E = mc^2$. At first, it might seem that we could simply go on recycling and reusing energy forever. If so, sustainability, meaning the ability to continue performing work, would be inevitable.

However, each time energy is used to perform work, some of its usefulness is lost. Once energy is used, before it can be used again, it must be reconcentrated, reorganized, and restored, and it takes energy to reconcentrate, reorganize, and restore energy. The energy used to reconcentrate, reorganize, and restore energy is simply no longer available to do anything else. It has lost its usefulness. This is the law of entropy: the tendency of all closed systems toward the ultimate degradation of matter and energy; a state of inert uniformity of component elements; an absence of structure, pattern, organization, or differentiation.[10] The desolate surfaces of the moon and Mars are systems as close to entropy as most of us have seen. Since this loss of useful energy is inevitable, it might seem that sustainability is impossible. And in fact, life on earth would not be sustainable without the daily inflow of solar energy that can be used to offset the usefulness of energy lost to entropy.

So what does this have to do with American agriculture? Industrial systems, including industrial agriculture, are very efficient in using and reusing the energy embodied in natural resources, but they do nothing to offset the inevitable loss of usefulness of energy due to entropy. Industrialization is driven by the economic motives of maximum profits and growth, and these economic benefits accrue to individuals and thus must accrue within individual lifetimes. It makes no economic sense to invest in restoring natural resources for the benefit of someone else of some future generation. Industrialization inevitably dissipates or uses up the physical energy embodied in natural resources, because industrialists have no incentive to renew or restore the resources from which they extract their productivity. Thus industrialization, by the logic and reason of the laws of science, quite simply is not sustainable.

Industrial farms, like other industrial organizations, are essentially resource-using systems. They use land, fertilizer, fuel, machinery, and people, but they do nothing to replace the energy that is inevitably lost when these resources are used to produce useful agricultural products. Industrial farmers don't use the solar energy from the sun to restore the productive capacities of their farms; instead, the crops and livestock produced in part from solar energy are sold off the farm to be used up elsewhere. It makes no economic sense for industrial farmers to maintain the productive capacity of the land if the benefits of doing so will accrue to someone of some future generation. An industrial agriculture accelerates the tendency toward entropy—it is not physically sustainable.

Industrialization not only uses up the natural resources required for sustainability, it also uses up the human resources. The law of entropy applies to social energy as well as physical energy. All human resources—labor, management, innovation, creativity—are products of social relationships. No person can be born or reach maturity without the help of other people who care about them personally, including their families, friends, neighbors, and communities. All organizations, including farms and businesses, depend on the ability of people to work together for a common purpose, which depends on the civility of the society in which they were raised.

Industrialization inevitably disperses and disorganizes *social* energy because it weakens personal relationships. Maximum economic efficiency requires that people relate to each other impartially, which means impersonally. People must compete rather than cooperate if market economies are to work efficiently. When family members work away from home to increase their productivity, they have less time and energy to spend together, and personal relationships are threatened. When people shop in another town rather than buy locally to save money, personal relationships among community members suffer from neglect. Industrialization inevitably devalues personal relationships and disconnects people, thus dissipating social energy. There are no economic incentives for industrialists to invest in renewing or restoring personal relationships within families or communities for the long-run benefit of society.

The industrialization of American agriculture, in fact, has torn the fabric of rural society apart. The specialization and mechanization of American agriculture has resulted in consolidation of farmland into larger and fewer farms, meaning fewer farm families. It takes people, not just production, to support rural communities—to buy feed, fuel, clothes, and haircuts on Main Street, to support local schools, churches, and other public services. Some farming communities have become so desperate they will grasp at any opportunity for survival. Unfortunately, outside investors see rural areas, with their open spaces and sparse population, as ideal places for things other communities don't want, such as prisons, urban landfills, toxic waste incinerators, or giant contract confinement animal feeding operations. Ethanol factories are just the latest "economic opportunity" to join this list. Such enterprises create economic benefits for a few but at the expense of the many—those who live downstream or downwind. The industrialization of agriculture inevitably creates conflicts and degrades relationships within rural communities. Industrial agriculture accelerates the depletion of social energy—it is not socially sustainable.

Economies are simply the means by which we deal with relationships among people and between people and the natural environment in complex societies. There are simply too many of us to barter with each other and to produce our own food, clothing, and shelter. Economies actually produce nothing; they simply transform physical energy and social energy into forms that can be traded or exchanged in impersonal marketplaces. All economic capital, meaning anything capable of producing anything of economic value, is extracted from either "natural capital" or "social capital." Industrial agriculture extracts its economic resources from the earth and from society; it uses up the fertility of the farmland and the productive capacities of rural people. Thus, when all the physical and social energy of rural areas has been extracted and exploited, industrial agriculture will have nothing left to support it economically. Industrial agriculture inevitably tends toward entropy—it is not economically sustainable.

A sustainable economy must be fundamentally different from the industrial economy of today. A sustainable economy must be based on the paradigm of living systems. Living things are self-making, self-renewing, reproductive, and regenerative.[11] Living plants have the natural capacity to capture, organize, concentrate, and store solar energy, both to support other living organisms and to offset the energy that is inevitably lost to entropy. Living things have a natural propensity to reproduce their species, and thus to renew and regenerate energy. Humans, for example, devote large amounts of time and energy to raising families, with very little economic incentive to do so. Obviously, an individual life is not sustainable because every living thing eventually dies. But communities and societies of living individuals clearly have the capacity and natural propensity to be highly productive while devoting a significant part of their life's energy to renewing the ecological and social capital needed to sustain economic capital.

The only renewable source of energy for the future is solar energy. Wind, water, and photovoltaic cells are the most promising sources of energy for the future. Even a society that relies on renewable solar energy, however, must continue to invest energy in renewing and regenerating material energy resources for the future. Windmills, water generators, and solar cells are made of physical materials that eventually wear out and must be replaced. No society will ever be sustainable unless its members are willing to make investments not just for themselves but also for the benefit of the future of humanity.

The logical alternative to the energy-using industrial agriculture of today is an energy-renewing sustainable agriculture. The sustainable agriculture movement emerged in the U.S. during the 1980s from growing concerns about declining farm profitability, environmental impacts of agrochemicals, and the viability of rural communities. The sustainable agriculture movement includes farmers who identify with organic, biodynamic, holistic, bio-intensive, biological, ecological, and permaculture, as well as many who claim no identification other than traditional family farmer. These farmers and their customers share a common commitment to creating an agriculture that is capable of maintaining its productivity and value to society indefinitely.

Sustainable farms must be ecologically sound, socially responsible, and economically viable. A farm that degrades the productivity of the land or pollutes its natural environment cannot sustain its productivity. A farm that fails to meet the needs of a society—not only as consumers but also as producers and citizens—cannot be sustained over time by that society. And a farm that is not profitable, at least over time, is not economically sustainable, regardless of its ecological and social performance. A sustainable farm ultimately must rely on renewable solar and social energy for its economic productivity.

Sustainable agriculture embraces the historic principles of organic farming. Sir Albert Howard, a pioneer of organic farming, began his book *An Agricultural Testament* with the assertion, "The maintenance of the fertility of the soil is the first condition of any permanent system of agriculture."[12] He contrasted the permanent agriculture of the Orient with the agricultural decline that led to the fall of Rome. He concluded, "The farmers of the West are repeating the mistakes made by Imperial Rome." J. I. Rodale, another prominent proponent of organic farming, wrote, "The *organiculturist* farmer must realize that in him is placed a sacred trust. . . . As a patriotic duty, he assumes an obligation to preserve the fertility of the soil, a precious heritage that he must pass on, undefiled and even enriched, to subsequent generations."[13]

Rudolph Steiner, the founder of biodynamic farming, defined an organic farm as a living system, as an organism whose health and productivity depended on healthy relationships among its ecological, social, economic, and spiritual dimensions. He wrote, "A farm is healthy only as much as it becomes an organism in itself—an individualized, diverse ecosystem guided by the farmer, standing in living interaction with the larger ecological, social, economic, and spiritual realities of which it is part."[14] To Steiner, organic farming was about the farmer becoming an integral part of a natural, living, productive, regenerative system.

Sustainable farmers rely on green plants to capture and store solar energy and to regenerate the organic matter and natural productivity of the soil. They use crop rotations, cover crops, intercropping, managed grazing, and integrated crop and livestock systems to manage pests and to maintain the natural fertility of their soils. Sustainable farmers reflect a sense of ethical and moral commitment to preserve the productivity of their land—to leave it as good as or better than they found it. Many of today's industrial organic producers have adopted large-scale, specialized, standardized systems, but sustainable organic farmers have remained committed to creating a permanent agriculture capable of supporting a permanent society.

Even though agriculture cannot generate enough renewable energy to replace fossil fuels, a shift from industrial to sustainable agriculture could reduce the dependence of food production on fossil energy. Shifting to a vegetarian diet has been suggested as a means of reducing energy use in agriculture, since most food crops are net energy producers and livestock are net energy users. A vegetarian diet might cut the food energy input/output ratio in half, since today's meat animals are inefficient converters of fossil energy into food energy. But the energy used and lost in processing and distribution would still leave a deficit of about five kilocalories of fossil energy to each kilocalorie of food energy.[15] In addition, pastures and forages are large net energy producers, accounting for about of 80 percent of total solar energy captured by agriculture. Meat is also a major source of food protein, much of which is produced from pastures and forages. Most pastures and forages are grown on land unsuitable for food crops, and pastures and forages cannot be digested by humans. Thus, a vegetarian diet would sacrifice most of the solar energy currently captured by agriculture.

Significant fossil energy savings for sustainably produced livestock and poultry might well be achievable without critical reductions in animal protein. Shifting from confinement livestock feeding to forage-and grass-based operations would be a more logical means of reducing the energy used in animal agriculture. A shift to grass-based systems could save an estimated 35 percent of the total energy now used in beef, dairy, and lamb production.[16] Sustainable grass-based livestock systems, utilizing management-intensive grazing, are capable of producing from 50 to 100 percent more protein per acre than a conventional pasture or forage system, while using less fertilizer, pesticides, and fuel. Free-range and pasture-based pork and poultry operations also are far more energy efficient than confinement feeding operations. In addition, hogs and chickens are natural scavengers and thus could get a significant portion of their diets from biological waste products. The elimination of confinement animal feeding operations also would result in significant social, environmental, and diet-related benefits in addition to energy savings.

Sustainable crop production practices could reduce agricultural energy use even farther. For example, recent research, based on more than twenty years of data, indicates that shifting from conventional to organic farming practices could save as much as 30 percent of the fossil energy used in cropping systems without reducing total production.[17] Changes in food processing and distribution, such as increased use of raw and minimally processed foods, more meals prepared at home, and a shift to more community-based local food systems, could increase the efficiency of energy use in food marketing by comparable amounts.

Although no comprehensive studies have been done, shifting to a more sustainable agricultural system using currently available methods and technologies probably could cut total fossil energy use in agriculture as much as one-half, resulting in a savings equivalent to about 3 percent of total U.S. fossil energy use. Similar efficiencies in processing and distribution could save an additional 6 percent or so in fossil energy use, but would still leave total food production with an 8 percent fossil energy deficit. It seems unlikely that agriculture will ever be able to produce more energy than will be needed to meet the increasing food and fiber needs of people. But a sustainable agriculture eventually could make food production fossil-energy independent—by supplementing solar energy captured by crops with energy from windmills, falling water, and photovoltaic cells. Such goals are fundamentally incompatible with industrial agriculture, which will continue to rely on nonrenewable resources for maximum short-run productivity until those resources are gone.

Many conventional farmers have considered sustainable agriculture to be a niche market, okay for a few small "fringe farmers" but not for "mainstream agriculture." Today, however, sustainable agriculture, by its various names, is showing signs of becoming the new agricultural mainstream. Organic foods first brought widespread attention to sustainable agriculture when organic food sales grew by more than 20 percent per year throughout the 1990s and early 2000s. In spite of this rapid growth, organic foods still only accounted for about 2 to 3 percent of total food sales in 2005. However, the potential market for sustainable local foods now appears to be far larger.

Various studies and surveys indicate that as many as one-third of American consumers have core values consistent with the principles of sustainability.[18] Thus, sustainable farmers today have an opportunity to help create a new food production and marketing mainstream simply by developing relationships with like-minded customers who share their same core values. In addition, sustainable farmers are finding new allies as more independent food processors, distributors, and retailers realize they face the same kinds of challenges and have the same kinds of opportunities as do independent family farmers. Sustainable agriculture has moved beyond farmers' markets, CSAs, and roadside stands—even though these markets will continue to grow—and is moving into the higher-volume retail food markets. Independently owned and operated supermarkets and restaurants create new opportunities for those on small and moderate-sized farms to work together to help create a new sustainable American food system.

In some cases, the initiative for creating the new food system is coming from the retail level. For example, New Seasons Market is one of the fastest-growing retail food chains in Portland, Oregon, currently operating seven stores with plans to operate at least nine. As Brian Rohter, cofounder and president writes, "Three families and about fifty of our friends decided in late 1999 that we wanted to create a business that we could be proud of—a company that had a true commitment to its community, to promoting sustainable agriculture and to maintaining a progressive workplace."[19] New Seasons markets look like other modern food supermarkets, with delis, bakeries, and other amenities Americans have come to expect. Once inside the store, the most noticeable difference is that virtually every item in the store is labeled with respect to origin, and there is an organic and conventional option for nearly every food item. In 2005, New Seasons started a new "Home Grown" program to promote items produced in Washington, Oregon, and northern California and to buy as many food items as possible from this region.

Sometimes the initiative has come from farmers. For example, Good Natured Family Farms is a cooperative made up of thirty-some farmers in southeastern Kansas and southwestern Missouri. Diana Endicott, who farms with her husband, Mel, was the moving force in gaining access to Kansas City supermarkets. Today they market most of their products through Hen House Markets, a thirteen-store supermarket chain operated by Ball Foods Inc., a family corporation with a long history of community connections in Kansas City. The cooperative owns and manages their own brand, which broke into the market with premium, locally grown beef but now includes an expanding line of food products with chicken, eggs, sausages, and milk, and other products are in various stages of development.

Their Web site states, "We have three goals: Support local farmers by providing them with a market for the food they raise, provide our customers with fresh, natural foods raised humanely, without hormones or sub-therapeutic antibiotics, and raise our beef, chicken, eggs, and milk in a manner which protects and conserves the precious resources upon which they rely."[20] Diana also serves as marketing liaison between Ball Foods and a number of other local growers who provide a wide range of local products. Their "Buy Fresh, Buy Local" campaign has resulted in several consecutive years of 35 percent increases in sales of "local products."

At times, the initiative has come from nonprofit organizations with an interest in promoting the common interests of farmers, processors, retailers, and consumers. FoodTrust of Prince Edward Island represents farmers and participants in all other aspects of the food system in the province. More than one hundred A&P Canada supermarkets offer FoodTrust potatoes, spices, and condiments to their customers. FoodTrust farmers are expected to meet a stringent set of standards for food safety and environmental stewardship, and even standards of social responsibility. Their Web site states, "Today, people are increasingly concerned about where their food comes from and how it is grown. FoodTrust responds to this concern by providing an important link between you and a group of dedicated farmers and growers who produce and harvest high quality, safe and wholesome foods."[21]

These are just three of many notable examples of people who are committed to creating a new sustainable local food system. Others who are leading the way include restaurants, both upscale and family diners; institutions such as public schools, universities, and prisons; and a wide variety of farmers' cooperative ventures. For example, more than five hundred public school districts and ninety-five colleges and universities have active programs to provide U.S. students with locally grown foods.[22]

The future of agriculture is in producing food for people,

not fuel for automobiles. There is nothing wrong with farmers reaping the short-term benefits of government subsidies to provide liquid energy for an energy-addicted society, as long as they don't lose sight of their future. Production of ethanol and biodiesel is just another industrial development strategy that ultimately will result in the further exploitation of farmland, farmers, and people of rural communities. Ultimately, the large agribusinesses, such as Archer Daniels Midland and Continental, and the large energy companies will control ethanol and biodiesel production and distribution. Those few farmers who invested in the initial energy co-ops may make money. Eventually, however, they either will be bought out or driven out of business by larger corporate competitors. Energy crops of the future will be produced under comprehensive corporate contracts, with producers receiving just enough money to put in another crop. Young people will continue to leave rural areas and rural communities will continue to wither and die.

American society eventually must break its addiction to fossil energy, and farmers must break their addiction to government subsidies and corporate domination. As fossil energy becomes increasingly scarce and expensive, Americans will turn to alternatives other than biofuels to meet their energy needs. Petroleum from tar sands, coal, and oil shale are all much more energy efficient than are biofuels from agriculture. Energy from wind, water, and photovoltaic cells are all far more energy efficient than is energy from biofuels. Biofuels can simply be brought on line much more quickly and with a much smaller initial investment, particularly with the promise of continuing government subsidies. Once the more energy-efficient alternative fuels come on line, rural communities will be left with abandoned ethanol and biodiesel plants and the environmental mess to clean up.

Eventually Americans will come to realize that a fossil-energy-dependent agriculture cannot replace fossil energy. As fossil energy becomes increasingly scarce and expensive, Americans eventually also will realize that the highest priority for agriculture must be to produce more food with less nonrenewable energy. America has perhaps a fifty-year window of opportunity to develop a sustainable, fossil-energy-independent food system.

It can be done. Many organic and sustainable farmers today produce just as much per acre as their industrial counterparts; they just have to put more of themselves into the production process. It will not take more land but it will take more farmers—more thinking, innovative, creative, caring farmers. It will also take more caring food consumers who are willing to pay the full ecological and social costs of sustainable food production. And it will make more independent food processors and distributors willing to work with farmers and consumers to build a more sustainable food system. And all of this will take time. So now is the time to get serious about creating the kind of agriculture that America must have to survive, after fossil energy.

⋮ Paper presented at the Iowa Farmers Union Annual Conference, Ankeny, Iowa, August 25–26, 2006.

Notes

chapter 1. *Crisis and Opportunity*

1. The "agricultural establishment," as used here, refers to the USDA, major agricultural universities, major commodity organizations, and some of the general farm organizations, such as the Farm Bureau Federation.

2. Steven Blank, *The End of Agriculture in the American Portfolio* (Westport CT: Quorum Books, 1999).

3. *Webster's New Collegiate Dictionary*, 8th ed., s.v. "crisis."

4. Adam Smith, *Wealth of Nations* (1776; Amherst NY: Prometheus Books, 1991).

5. The stories of many of these new American farmers have now been documented in a variety of formats. For example, see "Sustainable Agriculture Network" in *The New American Farmer*, 2nd ed., ed. Valerie Berton (Beltsville MD: U.S. Department of Agriculture, 2005), also available at http://www.sare.org/publications/ naf (accessed June 2006); and "1,000 Stories of Regenerative Agriculture," The New Farm (Rodale Institute), http://www .newfarm.org/archive/1000_stories/1000_stories.shtml (accessed June 2006).

6. Howard Elitzak, *Food Cost Review, 1996* (Washington DC: U.S. Department of Agriculture, 1997).

7. For example, Dennis Avery, *The Environmental Triumph of High-Yield Farming* (Washington DC: Hudson Institute, 1995).

8. Thomas Malthus, *An Essay on the Principles of Population* (1789; London: Macmillan, 1966).

chapter 2. *Why We Should Stop Promoting Industrial Agriculture*

1. Peter Drucker, *Post-Capitalist Society* (New York: HarperBusiness, 1994), 1.

2. U.S. Department of Commerce, *1992 Census of Agriculture* (Washington DC: U.S. Census Bureau, 1994).

3. Howard Elitzak, "Food Marketing Costs Rose Little 1992," *Food Cost Review* (Washington DC: U.S. Department of Agriculture, 1993).

4. Elitzak, *Food Cost Review*; U.S. Department of Agriculture, *1993 Farm Cost and Return Survey* (Washington DC: National Agricultural Statistical Service, 1994).

5. David Pimentel and Mario Giampietro, *Food, Land, Population and the U.S. Economy* (Washington DC: Carrying Capacity Network, 1994), also available at http://dieoff.org/page55.htm (accessed June 2006).

6. Alvin Toffler, *PowerShift* (New York: Bantam Books, 1990).

7. Toffler, *PowerShift*, 9.

8. Toffler, *PowerShift*, 238.

9. Toffler, *PowerShift*, 91.

10. Peter Drucker, *The New Realities* (New York: Harper and Row, 1989), 173.

11. Robert B. Reich, *The Work of Nations* (New York: Vintage Books, 1992), 174.

12. Joel Barker, *Paradigms: The Business of Discovering the Future* (New York: HarperBusiness, 1993).

13. Barker, *Paradigms*.

14. Drucker, *Post-Capitalist Society*, 210.

CHAPTER 3. *Corporate Ag and Family Farms*

1. U.S. Census Bureau, "Census of Population and Housing," http://www.census.gov/prod/www/abs/decennial/index.htm (accessed June 2006).

2. Economic Research Service, "U.S. Farm Income Data," historical documents available online at http://www.ers.usda.gov/Data/FarmIncome/finfidmu.htm (accessed June 2006).

3. Economic Research Service, *Structural and Financial Characteristics of U.S. Farms, 1995* (Washington DC: U.S. Department of Agriculture, 1996).

4. See, for example, the Web sites "Sustainable Agriculture Network" and "1,000 Stories of Regenerative Agriculture."

5. This section has been condensed from the original version of this paper to reduce duplication of material in the title essay of this book.

6. Wendell Berry, "Nature as Measure," in *What Are People For?* (New York: North Point Press, 1997), 206–7.

CHAPTER 4. *Corporatization of America*

1. Stephen Covey, *Seven Habits of Highly Effective People* (New York: Simon and Schuster, 1989).

2. *Webster's New Collegiate Dictionary*, 8th ed., s.v. "corporatism."

3. *Microsoft Encarta Encyclopedia 2003* (Redmond WA: Microsoft Corporation, 1993–2002), s.v. "Industrial Revolution."

4. *Microsoft Encarta*, s.v. "Progressive Movement."

CHAPTER 5. *Rediscovering Agriculture*

1. Vaclav Havel, *Disturbing the Peace* (New York: Vintage Books, 1991), 181, 182, 188.

2. This section describing the new sustainable farmers has been shortened from the original version of this paper to reduce duplication of material from previous chapters.

3. See the Web sites "Sustainable Agriculture Network" and "1,000 Stories of Regenerative Agriculture."

CHAPTER 6. *Farming in Harmony*

1. William James, *The Varieties of Religious Experience* (1902; New York: Routledge, 2002).

2. Joel Salatin, *Pastured Poultry Profits* (Swoope VA: Polyface Inc., 1996).

3. Berry, "Nature as Measure," 210.

CHAPTER 7. *Reclaiming the Sacred*

1. Fritjof Capra, *Tao of Physics* (Boulder CO: Shambhala Publications, 1983), 22.

2. E. F. Schumacher, *Small Is Beautiful* (New York: Harper and Row, 1975).

3. Dee Hock, "The Chaordic Organization: Out of Control and Into Order," *World Business Academy Perspectives* 9, no. 1 (1995): 5–21.

4. Toffler, *PowerShift*; Vaclav Havel, acceptance speech upon receiving the Philadelphia Liberty Medal, July 4, 1994, http://old .hrad.cz/president/Havel/speeches/1994/0407_uk.html (accessed June 2006); Tom Peters, *The Pursuit of wow!* (New York: Vantage Books, 1994); Drucker, *New Realities*; John Naisbitt and Patricia Aburdene, *Megatrends 2000* (New York: Avon Books, 1990); Reich, *Work of Nations*.

5. Rachel Carson, *Silent Spring* (Boston: Houghton Mifflin, 1962).

6. Rushworth W. Kidder, *Reinventing the Future* (Cambridge MA: Christian Science Publishing Society, MIT Press, 1989).

7. James, *Varieties of Religious Experience*.

8. Elizabeth Roberts and Elias Amidon, *Earth Prayers from Around the World* (New York: Harper-Collins, 1991), 10.

9. Fritjof Capra, *The Turning Point: Science, Society, and the Rising Culture* (New York: Simon and Shuster, 1982), 131.

10. Paul Pearsall, *The Pleasure Prescription* (Alameda CA: Hunter House Publishing, 1996), 121.

11. Roberts and Amidon, *Earth Prayers*, 62.

CHAPTER 9. *Foundational Principles*

1. *Webster's New Collegiate Dictionary*, 8th ed., s.v. "foundation."

2. Selina Gaye, *The Great World's Farm: Some Accounts of Nature's Crops and How They Are Grown* (New York: Macmillan, 1900).

3. Gaye, *Great World's Farm*, 61.

4. William A. Albrecht, "Protein Deficiencies . . . through Soil Deficiencies," *Let's Live Magazine*, December 1952, also available at http://www.soilandhealth.org/copyform.asp?bookcode=010143let slive1953 (accessed June 2006).

5. *Webster's New Collegiate Dictionary*, 8th ed., s.v. "stewardship."

6. R. B. Beverly and S. L. Ott, "Spiritual Stewardship: A Conceptual Framework for Teaching Ethics in Agriculture," *Agronomy Education* 18, no. 2 (1989): 121–24.

7. James, *Varieties of Religious Experience*.

8. Aldo Leopold, *A Sand County Almanac* (1949; New York: Ballantine Books, 1970), 262.

CHAPTER 10. *Economics of Sustainable Farming*

1. Alan Savory, *Holistic Resource Management* (Covelo CA: Island Press, 1988); Rudolph Steiner, *Spiritual Foundations for the Renewal of Agriculture*, ed. M. Gardner (1924; Junction City OR: Biodynamic Farming and Gardening Association of the USA, 1993), also available at http://www.biodynamics.com/index.html (accessed June 2006); Permaculture Research Institute, "What Is Permaculture?" http://www.permaculture.org.au/ (accessed June 2006); and Organic Farming Research Foundation, "What Is Organic Agriculture?" http://www.ofrf.org/general/about_organic/index.html (accessed June 2006).

CHAPTER 11. *Renaissance of Rural America*

1. Robert Pool, "Science Literacy: The Enemy Is Us," *Science* 251, no. 4991 (1991): 267.

2. Elizabeth Culotta, "Science's 20 Greatest Hits Take Their Lumps," *Science* 251, no. 4999 (1991): 1308.

3. Toffler, *PowerShift*, xvii.

4. Toffler, *PowerShift*, 9.

5. Toffler, *PowerShift*, 238, 91.

6. As quoted in Peter M. Senge, *The Fifth Discipline* (New York: Doubleday, 1990), 366.

7. Senge, *Fifth Discipline*, 367.

8. Drucker, *New Realities*, 173.

9. Drucker, *New Realities*, 259.

10. Drucker, *New Realities*, 260.

11. Reich, *Work of Nations*.

12. Naisbitt and Aburdene, *Megatrends 2000*.

13. National Science Board, *Science Indicators 1976* (Washington DC: National Science Foundation, 1977), 91–128.

14. Naisbitt and Aburdene, *Megatrends 2000*, 329.

15. Naisbitt and Aburdene, *Megatrends 2000*, 332.

16. Drucker, *New Realities*, 259.

17. This section draws from "Sustainable Agriculture and Quality of Life," an unpublished USDA Sustainable Agriculture Research and Education Report, prepared by the Sustainable Agriculture Quality of Life Task Force, from the files of the author.

18. This section was revised significantly from the original manuscript to reduce duplication of material from other essays.

CHAPTER 12. *Walking the Talk*

1. The five conferences are the Upper Midwest Organic Growers, typically held in Wisconsin; the Pennsylvania Association for Sustainable Agriculture; the Eco-farming and Bioneers conferences in California; and the Northeast Organic Farming Association in New York. By 2005 attendance at each of these conferences ranged from 1,500 to 4,500 people.

2. I have condensed my description of the new American farmers from the original paper in order to reduce repetition of material from previous chapters. The stories of many of these New American Farmers have now been documented in a variety of formats. See "Sustainable Agriculture Network" and "1,000 Stories of Regenerative Agriculture."

3. Based on surveys conducted by *Growing for Market*, a monthly newsletter of Fairplain Publications, Lawrence KS, http://www.growingformarket.com/ (accessed June 2006).

CHAPTER 13. *Survival Strategies*

1. Sources of statistics cited in this section were familiar to the audience for which this paper was written. Statistics can be validated by referring to the sources cited in chapter 3: U.S. Census Bureau, *Census of Population and Housing*, historical census documents available online at http://www.census.gov/prod/www/abs/decennial/index.htm (accessed June 2006); Economic Research Service, U.S. Farm Income Data, historical documents available at http://www.ers.usda.gov/Data/FarmIncome/finfidmu.htm (accessed June 2006); and Economic Research Service, *Structural and Financial Characteristics of U.S. Farms, 1995* (Washington DC: U.S. Department of Agriculture, 1996).

2. Willard W. Cochrane, "A Food and Agricultural Policy for the 21st Century," Agobservatory Library, 1999, http://www.agobservatory.org/library.cfm?RefID=29732 (accessed June 2006).

CHAPTER 14. *Marketing in the Niches*

1. Sections related to sustainability and industrialization are condensed from the original paper in order to reduce duplication of material in previous chapters.
2. Elitzak, "Food Marketing Costs Rose Little 1992."
3. Robert J. Kriegel and Louis Patler, *If It Ain't Broke—Break It! And Other Unconventional Wisdom for a Changing Business World* (New York: Warner Books, 1991).

CHAPTER 15. *Local Organic Farms*

1. USDA Economic Research Service, "A County-Level Measure on Urban Influence," *Rural Development Perspectives* 12 no. 2 (1997).
2. U.S. Environmental Protection Agency, "Managing Non-Point Source Pollution from Agriculture" 2004, http://www.epa.gov/owow/nps/facts/point6.htm (accessed June 2006); Donald M. Anderson, "Toxic Algal Blooms and Red Tides: A Global Perspective," in *Red Tides: Biology, Environmental Science, and Toxicology*, ed. Tomotoshi Okaichi, Donald M. Anderson, and Takahisa Nemoto (New York: Elsevier, 1989), 11–16.
3. Organic Farming Research Foundation, *Third Biennial National Organic Farmer's Survey* (Santa Cruz CA: Organic Farming Research Foundation, 1999).
4. John Ikerd, "The Architecture of Organic Production," *Proceedings, Inaugural National Organics Conference 2001, Sidney, Australia,* 2001, 27–28.
5. Hartman Report, "Food and the Environment—A Consumer's Perspective," 1999, http://www.hartman-group.com/products/reportnatsens.html (accessed June 2006).
6. The description of the new food culture has been condensed from the original paper to reduce duplication of material in chapter 19, "The New American Food System."
7. Eliot Coleman, "Can Organics Save the Family Farm?" *The Rake Magazine*, September 2004, also available at http://www.rakemag.com (accessed June 2006).
8. Steiner, *Spiritual Foundations.*
9. Albrecht, "Protein Deficiencies."

10. J. I. Rodale, "The Organiculturist's Creed," in *The Organic Front* (Emmaus PA: Rodale Press, 1948), also available at http://www .soilandhealth.org/copyform.asp?bookcode=010133 (accessed June 2006).
11. Sir Albert Howard, *An Agricultural Testament* (Oxford: Oxford University Press, 1940), also available in Small Farms Library, http://journeytoforever.org/farm_library/howardAT/ATtoc.html (accessed June 2006).
12. G. T. Wrench, "The Second Path," in *The Restoration of the Peasantries* (London: C. W. Daniel Company, 1939), also available in Small Farms Library, http://journeytoforever.org/farm_library .html#Wrench_Rest (accessed June 2006).
13. *Webster's Third New International Dictionary, Unabridged* (1993), s.v. "entropy."
14. Berry, "Nature as Measure," 210.
15. Ohio State University, "Cluster Development," Community Development Fact Sheet CDFS-1270–99, http://ohioline.osu.edu/ cd-fact/1270.html (accessed December 2004).

CHAPTER 16. *Triple Bottom Line*

1. *Webster's Eighth New Collegiate Dictionary*, 1973 ed., s.v. "crisis."
2. Fred Kirschenmann, Steve Stevenson, Fred Buttel, Tom Lyson, and Mike Duffy, "A White Paper for Agriculture of the Middle Project" 2003, Agriculture of the Middle, http://www.agofthemiddle .org (accessed June 2006).
3. National Agricultural Statistical Service, *2002 Census of U.S. Agriculture* (Washington DC: U.S. Department of Agriculture, 2003).
4. Willard W. Cochrane, "A Food and Agricultural Policy for the 21st Century," Agobservatory Library, 1999, http://www.agobservatory .org/library.cfm?RefID=29732 (accessed June 2006).
5. The discussion of ecological and social concerns in this section was condensed from the original paper to reduce duplication of material from the previous chapter.
6. For examples of scientific evidence, see Kendall M. Thu and E. Paul Durrenberger, *Pigs, Profits, and Rural Communities* (Albany: State University of New York Press, 1998).
7. Economic Research Service, "The U.S. Ag Trade Balance ... More

than Just a Number," *Amber Waves*, February 2004, http://www
.ers.usda.gov/amberwaves/February04/Features/USTradeBalance
.htm (accessed June 2006).

8. Steven Blank, *The End of Agriculture in the American Portfolio*.

9. "The WTO Framework Agreement for Fairer Trade," August 1,
2004, Europa: Agriculture—International Trade Relations, http://
europa.eu.int/comm/agriculture/external/wto/index_en.htm (accessed June 2006).

10. John Elkington, *Cannibals with Forks: The Triple Bottom Line of 21st
Century Business* (Stony Creek CT: New Society Publishers, 1998).

11. Paul Hawken, Amory Lovins, and L. Hunter Lovins, *Natural Capitalism* (New York: Little, Brown, 1999).

12. Bob Willard, *The Sustainability Advantage: Seven Business Case Benefits of a Triple Bottom Line* (Gabriola Island BC: New Society Publishers, 2002).

13. Wayne Norman and Chris MacDonald, "Getting to the Bottom
of the 'Triple Bottom Line,'" *Business Ethics Quarterly*, April 2004,
243–62.

14. Ray Anderson, *Mid-Course Correction: Toward a Sustainable Enterprise: The Interface Model* (White River Junction VT: Chelsea Green
Publishers, 1998).

15. Fetzer Vineyards, "Environmental Philosophy," http://www
.fetzer.com/fetzer/wineries/philosophy.aspx (accessed June 2006).

16. New Seasons Market homepage, http://www.newseasonsmarket.
com (accessed June 2006).

17. Smith, *Wealth of Nations*, 83.

18. David Ricardo, *The Works and Correspondence of David Ricardo*,
ed. Piero Sraffa, vol. 1 (Cambridge: British Royal Society, 1951),
132.

19. Vilfredo Pareto, *Manual of Political Economy* (1906; New York:
Kelly, 1971).

20. Alfred Marshall, *Principles of Economics* (1890; London: Macmillan,
1946), 27.

21. J. R. Hicks and R. Allen, "A Reconsideration of the Theory of
Value," *Economica*, no. 1 (1934): 128.

22. William Hamilton, *Essays in Edinburgh Review* (Edinburgh: Edinburgh Review, 1829), 32.

23. Thomas Reid, *Works of Thomas Reid*, ed. William Hamilton (1863; Bristol, England: Thoemmes Press, 1994), 422.

CHAPTER 17. *Real Costs of Globalization*

1. World Trade Organization, Welcome to the WTO Website, http://www.wto.org/ (accessed June 2006).
2. *Webster's New Collegiate Dictionary*, 8th ed., s.v. "globalize."
3. NAFTA Secretariat, Welcome Page, http://www.nafta-sec-alena.org/DefaultSite/index_e.aspx (accessed June 2006); "The EU at a Glance," Europa: Gateway to European Union, http://europa.eu.int/abc/index_en.htm (accessed June 2006).
4. Ricardo, *Works and Correspondence*, 132.

CHAPTER 18. *Redirecting Government Policies*

1. U.S. Department of Agriculture, "Farm Security and Rural Investment Act of 2002," available at http://www.usda.gov/farmbill2002/ (accessed June 2006).
2. Economic Research Service, U.S. Department of Agriculture, "Government Payments and the Farm Sector," Farm and Commodity Policy (2003), available at http://www.ers.usda.gov/briefing/FarmPolicy/gov-pay.htm (accessed June 2006).
3. Economic Research Service, U.S. Department of Agriculture, "Outlook for U.S. Agricultural Trade" (2003), available at http://www.ers.usda.gov/publications/so/view.asp?f=trade/aes-bb/ (accessed June 2003).
4. Steven Blank, *The End of Agriculture in the American Portfolio*.
5. The remainder of this section of the paper has been condensed from the original paper, both to reduce details of the 2002 farm bill, which are no longer particularly relevant to farm policy, and to reduce duplication of discussions in other chapters.
6. Economic Research Service, "Government Payments and the Farm Sector."
7. Berry, "Nature as Measure," 206–7.
8. Sustainable Agriculture Network, *New American Farmer*.
9. Willard Cochrane, *The Curse of American Agricultural Abundance* (Lincoln: University of Nebraska Press, 2003).

CHAPTER 19. *New American Food System*

1. Robert Putnam, *Bowling Alone* (New York: Simon and Schuster, 2000).
2. Mary Hendrickson and William Heffernan, "Opening Spaces through Relocalization: Locating Potential Resistance in the Weaknesses of the Global Food System," *Sociologia Ruralis* 42, no. 4 (2002): 347–69.
3. Eric Schlosser, *Fast Food Nation* (Boston: Houghton Mifflin, 2001).
4. Paul Ray and Sherry Anderson, *The Cultural Creatives* (New York: Three Rivers Press, 2000).
5. Alex De Tocqueville, *Democracy in America* (1835; New York: Bantam Books, 2000).
6. Sustainable Agriculture Network, *New American Farmer*.
7. This section of the original paper has been edited to reduce duplication of material from other chapters.
8. The Hartman Report, "Food and the Environment—A Consumer's Perspective."
9. Chefs Collaborative, available at http://www.chefscollaborative.org/ (accessed February 2003).
10. Slow Food International, available at http://www.slowfood.com/ (accessed February 2003).
11. Corby Kummer, *The Pleasures of Slow Food* (San Francisco: Chronicle Books, 2002).
12. Malcolm Gladwell, *The Tipping Point* (Boston: Little, Brown, 2000).

CHAPTER 20. *Agriculture After Fossil Energy*

1. For a basic explanation of peak oil, see "The Community Solution," http://www.communitysolution.org/peakqanda.html (accessed December 2006).
2. Wikipedia, the free encyclopedia, "Hubbert Peak Theory," http://en.wikipedia.org/wiki/Hubbert_peak (accessed December 2006).
3. Richard Heinberg, *The Party's Over: Oil War and the Fate of Industrial Societies* (Gabriola Island BC: New Society Publishers), 152.
4. Heinberg, *The Party's Over*, 111–12.

5. Christopher Cook, "Business as Usual," *The American Prospect*, online edition, April 8, 2006, http://www.prospect.org/web/page .ww?section=root&name=ViewPrint&articleId=11322 (accessed December 2006).

6. From a presentation by David Pimentel, Cornell University, at "Local Solutions to Energy Dilemma," New York City, April 28–29, 2006. To account for increased energy use, this figure was revised from the earlier estimate of solar energy collected as two-thirds of fossil energy use, published in David and Marcia Pimentel, *Food, Energy, and Society* (Niwot CO: University Press of Colorado), 1996.

7. Pimentel, *Food, Energy, and Society*.

8. As reported by Alexei Barrionjevo, "It's Corn vs. Soybeans in a Bio-fuels Debate," *New York Times*, July 12, 2006.

9. Pimentel, *Food, Energy, and Society*.

10. For a more in-depth discussion of entropy, see John Ikerd, *Sustainable Capitalism: A Matter of Common Sense*, chap. 3 (Bloomfield CT: Kumarian Press Inc., 2005).

11. For a more in-depth discussion of living systems, see Ikerd, *Sustainable Capitalism*, ch. 5.

12. Howard, *An Agricultural Testament*.

13. J. I. Rodale, "The Organiculturist's Creed," in *The Organic Front* (Emmaus PA: Rodale Press, 1948), ch. 8, also available at http:// www.soilandhealth.org/copyform.asp?bookcode=010133 (accessed December 2006).

14. Steiner, *Spiritual Foundations*.

15. Pimentel, *Food, Energy, and Society*, 146.

16. Based on estimates in Pimentel, *Food, Energy, and Society*.

17. David Pimentel, Paul Hepperly, James Hanson, David Douds, and Rita Seidel, "Environmental, Energetic, and Economic Comparisons of Organic and Conventional Farming Systems," *BioScience* 55, no. 7 (2005): 573–82.

18. The Hartman Report estimates that two groups of consumers, the "new green mainstream" and the "true naturals," represent prime markets for natural foods and make up approximately 28 percent of all American consumers. See Hartman Report, "Food and the Environment—A Consumer's Perspective."

19. New Seasons Market homepage.
20. Good Natured Family Farms homepage, http://goodnatured.net/ (accessed December 2006).
21. FoodTrust Prince Edward Island, http://www.foodtrustpei.com/ about/ (accessed December 2006).
22. For more case studies, including institutional programs promoting local foods, see the Farm to School homepage, http://www .farmtoschool.org; Farm to College homepage, http://farmtocollege .org; and Agriculture of the Middle homepage, http://www .agofthemiddle.org/archives/2004/09/case_studies.html (accessed December 2006).

Ogallala: Water for a Dry Land, second edition
John Opie

Willard Cochrane and the American Family Farm
Richard A. Levins

*Down and Out on the Family Farm: Rural Rehabilitation
in the Great Plains, 1929–1945*
Michael Johnston Grant

*Raising a Stink: The Struggle over Factory Hog Farms
in Nebraska*
Carolyn Johnsen

*The Curse of American Agricultural Abundance:
A Sustainable Solution*
Willard W. Cochrane

Good Growing: Why Organic Farming Works
Leslie A. Duram

Roots of Change: Nebraska's New Agriculture
Mary Ridder

*Remaking American Communities: A Reference Guide
to Urban Sprawl*
Edited by David C. Soule
Foreword by Neal Peirce

*Remaking the North American Food System: Strategies
for Sustainability*
Edited by C. Clare Hinrichs and Thomas A. Lyson

*Crisis and Opportunity: Sustainability in
American Agriculture*
John E. Ikerd

Green Plans: Blueprint for a Sustainable Earth, revised
and updated
Huey D. Johnson